A Field Guide to
Coastal Wetland Plants of the
Northeastern United States

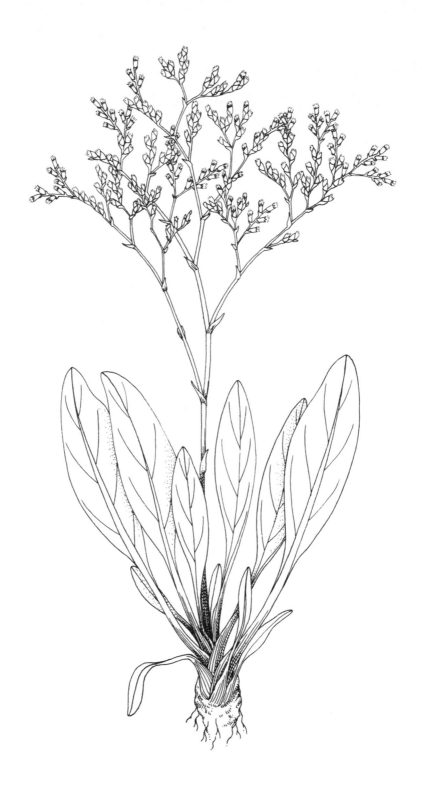

A Field Guide to
Coastal Wetland Plants of the
Northeastern United States

Ralph W. Tiner, Jr.

Drawings by Abigail Rorer

The University of Massachusetts Press

Amherst, 1987

Copyright © 1987 by Ralph W. Tiner, Jr.
Printed in the United States of America
Set in Linotron Sabon by G & S Typesetters, Inc.
Printed by Cushing-Malloy and bound by
John Dekker & Sons
Library of Congress Cataloging-in-Publication Data

Tiner, Ralph W.
    A field guide to coastal wetland plants of the
northeastern United States.

    Bibliography: p.
    Includes index.
    1. Wetland flora—Northeastern States—
Identification.   2. Coastal flora—Northeastern
States—Identification.   I. Title.
QK118.T56   1987   581.974   86-7076
ISBN 0-87023-537-0 (alk. paper)
ISBN 0-87023-538-9 (pbk.: alk. paper)

Dedicated to my mother
for her fortitude and
inspiration

Contents

Acknowledgments

I wish to extend my sincere gratitude and appreciation to those individuals who helped with the preparation of this book. Ken Metzler and Nels Barrett made available unpublished information on the distribution of coastal wetland plants in Connecticut's brackish and tidal fresh marshes; Ken also reviewed the draft of the plant identification keys and provided helpful suggestions. Dr. William Niering of Connecticut College and Dr. Paul Godfrey of the University of Massachusetts critically reviewed the draft manuscript and offered useful recommendations reflected in this book. The excellent plant illustrations were drawn by Abigail Rorer, whose contribution was invaluable to the utility of this field guide. Marian Rohman helped by providing specimens from the herbarium of the University of Massachusetts for the illustrations. The Maryland Department of Natural Resources, Tidewater Administration, provided funds from the National Oceanic and Atmospheric Administration, Office of Ocean and Coastal Resource Management, for most of the illustrations, and the efforts of David Burke in coordinating this are appreciated.

Numerous individuals provided information on places to observe wetlands: Janis Albright, Betsy Blair, James P. Browne, Carroll Curtis, Frank Dawson, Christy Foote-Smith, Sherrard C. Foster, Trisha Funk, David Hardin, James Kealy, Susan Latchum, Janet McMahon, Dr. W. W. Reynolds, Harry R. Tyler, Jr., and Mathilde P. Weingartner.

Special thanks are extended to Diana Doyle for her timely manuscript typing and her perseverance in preparing and proofreading the final copy. I am also indebted to my mother, Martha Tiner, for her coordination of the typing and to my wife, Barbara, for her support during this project.

I also wish to acknowledge the efforts of the University of Massachusetts Press, especially Bruce Wilcox and Barbara Werden, for preparing this book for publication.

A Field Guide to
Coastal Wetland Plants of the
Northeastern United States

Introduction

Coastal wetlands are among our most important fish and wildlife habitats. For example, over two-thirds of our recreationally and commercially important marine fishes depend on coastal marshes and associated waters for nursery and spawning grounds. Many waterfowl and wading birds also use these wetlands for feeding, resting, and nesting. Northeastern salt marshes are the primary overwintering area for black duck. Furbearers, such as muskrat and otter, make their homes in brackish and freshwater tidal marshes. Moreover, coastal wetlands serve as the farmlands for the estuarine and nearshore marine environments by producing millions of tons of organic matter annually. The most productive of our coastal marshes produce over 10 tons of organic matter per acre each year, which rivals our more productive cornfields. The organic matter comes mainly from the leaves and stems of herbaceous plants. Each fall, when these plants die back, their leaves and stems are gradually broken down into small fragments called *detritus*. This detritus provides food for a multitude of microorganisms (e.g., zooplankton), forage fishes (e.g., killifish, mullet, menhaden, and alewife), and grass shrimp, which in turn are the food for larger fishes, such as bluefish,

weakfish, and striped bass. These fishes are important food for humans, which completes the food chain link between the coastal wetlands and mankind. Coastal wetlands also offer us other values: (1) flood and storm damage protection by temporarily storing floodwaters and by buffering dry land from storm wave action, (2) water quality maintenance by removing sediment, nutrients, and other materials from flooding waters, (3) shellfish production, (4) recreation (e.g., waterfowl hunting, crabbing, fishing, nature photography, and bird watching), and (5) aesthetics.

Despite these natural values, coastal wetlands have been greatly mistreated in the past. Once considered as wastelands that bred mosquitoes, these wetlands were highly regarded as potential sites for development. Such development was thought to be beneficial because it helped reduce these public nuisances and provide more usable land. At that time, the relationship between these wetlands and coastal fisheries was poorly understood. After the Second World War, development of coastal wetlands accelerated as they were filled for industrial facilities and residential housing. While this was occurring, scientists studying coastal marshes found that they were productive areas essential to our coastal fisheries and other aquatic organisms. Coastal wetlands quickly became viewed as critical natural resources, worthy of special protection. Massachusetts was the first state to pass a law to protect these wetlands in 1963. By the mid-1970s, all northeastern states had enacted legislation regulating uses of coastal wetlands. The federal government also strengthened its role in conserving these areas through improved regulation of the River and Harbor Act of 1899 and by developing new regulations pursuant to the Federal Water Pollution Control Act of 1972 (later amended as the Clean Water Act of 1977). Today, coastal wetlands in the Northeast are receiving much better protection, and alternative uses are strictly controlled.

All of these laws define wetlands in part on the occurrence of plants adapted for life in wet areas. The state laws list numerous plants that are characteristic of tidal marshes. Plant identification is, therefore, an important step toward identifying coastal wetlands. For more than fifteen years, I have worked in the field of wetland mapping and protection and, to my continued surprise, a good field guide for identifying common northeastern coastal wetland plants has not been developed. The purpose of this book is to fill this void and to introduce readers to coastal wetland ecology. It is also intended to identify specific places where tidal wetlands can be visited and various sources of information about coastal wetlands. The focus of the book, however, remains on plant identification. This guide is designed primarily for the nontechnical person interested in learning to recognize common tidal wetland plants. It will be useful for representatives of federal and state environmental agencies, members of local conservation commissions, environmental consultants, landscape architects, students in botany and environmental sciences, interpretation specialists at parks and nature centers, and other individuals interested in coastal wetlands. This book is intended for primary use in coastal wetlands in the northeastern United States, from Maine through Maryland. Although the focus is on this area, most of the plants illustrated are also characteristic of southern coastal marshes. The guidebook should, therefore, be useful along the Atlantic coast from Maine and adjacent Canada to northern Florida, despite omissions of strictly southern species.

This book is arranged in five major sections: (1) Coastal Wetland Ecology: A General Overview, (2) Identification of Coastal Wetland Plants, (3) Wetland Plant Descriptions and Illustrations, (4) Places to Observe Coastal Wetlands, and (5) Sources of Other Information. In addition, a list of references used in preparing this book and a glossary of technical terms are provided.

Coastal Wetland Ecology:
A General Overview

Coastal wetlands are low-lying areas periodically flooded by tidal waters for varying lengths of time. They are associated with saltwater embayments and tidal rivers along the coastline. They generally include tidal flats, salt and brackish marshes, and tidal freshwater marshes and swamps. Other coastal wetlands are represented by intertidal rocky shores and beaches. Aquatic beds in permanently flooded areas are often associated with the intertidal coastal wetlands, either in pools and ponds within the marshes or in adjacent shallow open waters.

Flooding by tidal water is the common denominator of all coastal wetlands. It is the driving force that creates and maintains these habitats. Tidal flooding is highly variable, ranging from twice daily at low elevations to only a few times each year at the highest levels. The lowest portions of tidal flats may be almost continuously flooded and exposed to air only during extreme low tides. In most coastal wetlands, however, alternate flooding and exposure to air is the rule. Two general zones are recognized, based on the frequency of flooding: (1) regularly flooded zone and (2) irregularly flooded zone. The former zone is flooded at least once daily by tides, whereas

A

B

1. Coastal wetlands form behind barrier islands and beaches (A—Sea Isle City, New Jersey) and along tidal rivers (B—Connecticut River).

5

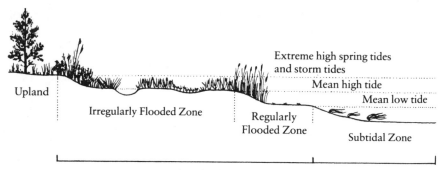

Upland

Irregularly Flooded Zone

Regularly
Flooded Zone

Extreme high spring tides
and storm tides

Mean high tide

Mean low tide

Subtidal Zone

Coastal Wetlands

Coastal Waters

2. Coastal wetlands can be divided into two zones, based on tidal hydrology: (1) regularly flooded zone and (2) irregularly flooded zone. The highest ocean-driven tides called "spring tides" occur during full and new moons; coastal storms can generate even higher tides, which may flood low-lying upland areas.

the latter zone is flooded less often and is generally exposed to air for long periods. Even when not flooded, the soils of coastal wetlands remain saturated at or near the surface. Plants adapted for life in flooded or saturated soils usually form extensive communities under these conditions.

Other factors, besides hydrology, are also operating to influence wetland vegetation. They include salinity, substrate type, temperature, biological competition, and human activities, such as ditching, pollution, and filling. Of these, salinity is perhaps the most important natural factor affecting plant growth.

Because coastal wetlands occupy positions between the open ocean and nontidal freshwater bodies, they largely exist in a zone of transition or flux where sea water intermixes with fresh water. This area of

mixing is called the *estuary*. An estuary includes both deepwater areas and contiguous intertidal wetlands. At the seaward end of the estuary, water salinity approaches sea strength (35 parts per thousand). As one moves farther upstream in coastal rivers, water salinity is diluted by increasing volumes of fresh water. This layer of salt water moves back and forth within the estuary as a wedge with the heaviest salt water at the bottom and a lens of fresh water at the surface. During periods of low freshwater discharge from rivers (e.g., late summer), the salt wedge penetrates to its farthest upstream point. By contrast, heavy river discharge such as spring runoff forces the salt wedge far downstream. In large coastal rivers with high freshwater discharge, eventually a point is reached where the water is strictly fresh, with no trace of ocean salts, yet the water

Table 1. Examples of mean and spring tidal ranges in the Northeast.

Location	Tidal Range (ft) Mean	Spring	Location	Tidal Range (ft) Mean	Spring
St. Croix River			Long Island Sound		
Calais, ME	20.0	22.8	Stonington, CT	2.7	3.2
Frenchman Bay			Mamaroneck, NY	7.3	8.6
Bar Harbor, ME	10.5	12.1	Great South Bay		
Hampton Harbor			Gilgo Heading, NY	1.1	1.3
Hampton, NH	8.3	9.5	Patchogue, NY	0.7	0.8
Merrimack River			Hudson River		
Newburyport, MA	7.8	9.0	Jersey City, NJ	4.4	5.3
			Tarrytown, NY	3.2	3.7
Cape Cod Bay			West Point, NY	2.7	3.1
Plymouth, MA	9.5	11.0	Albany, NY	4.6	5.0
Wellfleet, MA	10.0	11.6	Barnegat Bay		
Nantucket Sound			Mantoloking, NJ	0.5	0.6
Dennisport, MA	3.4	4.1	Barnegat Inlet, NJ	3.1	3.8
Falmouth Heights, MA	1.3	1.6	Delaware River		
Narragansett Bay			New Castle, DE	5.6	6.0
Newport, RI	3.5	4.4	Philadelphia, PA	6.0	6.3
			Trenton, NJ	6.8	7.1
Connecticut River			Chesapeake Bay		
Saybrook Jetty, CT	3.5	4.2	Kent Island, MD	1.2	1.4
East Haddam, CT	2.9	3.5			
Hartford, CT	1.9	2.3			

Note: Tidal ranges differ greatly throughout the region and even within the same coastal waterbody.

levels are still subject to rises and falls with the changing tides. Here, tidal waters moving upstream on the rising tide form a barrier that impedes much of the downstream flow of fresh water. This damming effect forces water levels to rise with the flooding tide. On the ebb or falling tide, the barrier is gradually removed, allowing the fresh water to flow freely downstream. Upstream beyond this area, the river is not influenced by the tides, and this begins the nontidal reaches of the river.

Under the variety of flooding, salinity, and other conditions, six major types of coastal wetlands are formed: (1) rocky shores, (2) tidal flats, (3) salt marshes, (4) brackish marshes, (5) tidal fresh marshes, and (6) tidal swamps. Shrub wetlands may be locally important in estuaries but are much less abundant than the other types. Each major coastal wetland type is generally described below. Aquatic beds are also discussed because they are closely associated with coastal wetlands.

Rocky Shores

Rocky shores dominate the coast of northern New England and can be found locally in southern New

SUMMER

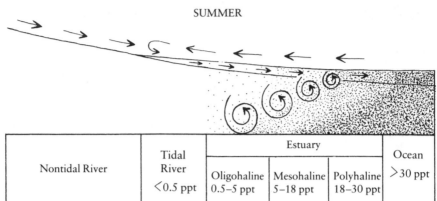

Nontidal River	Tidal River <0.5 ppt	Estuary			Ocean >30 ppt
		Oligohaline 0.5–5 ppt	Mesohaline 5–18 ppt	Polyhaline 18–30 ppt	

SPRING

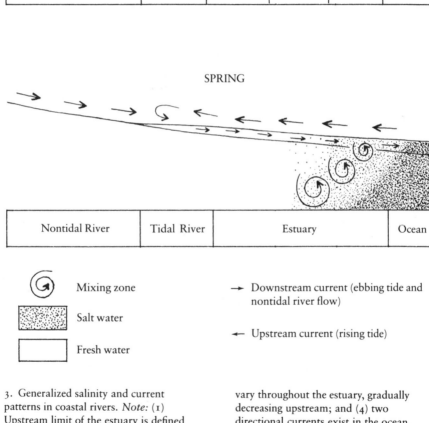

Nontidal River	Tidal River	Estuary	Ocean

⊙ Mixing zone

▦ Salt water

☐ Fresh water

→ Downstream current (ebbing tide and nontidal river flow)

← Upstream current (rising tide)

3. Generalized salinity and current patterns in coastal rivers. *Note:* (1) Upstream limit of the estuary is defined by maximum penetration of measureable sea water; (2) the position of the mixing zone changes seasonally; (3) salinities vary throughout the estuary, gradually decreasing upstream; and (4) two directional currents exist in the ocean, estuary, and tidal river due to the tides, whereas flow of nontidal river is one directional (downstream).

Table 2. Wetland types referred to in this book, with their corresponding technical classifications according to the U.S. Fish and Wildlife Service. This classification system was primarily developed for mapping the nation's wetlands.

Coastal Wetland Type Used in this Book	U.S. Fish and Wildlife Service Classification			
	System(s)	Subsystem	Class	Water Regime(s)
Rocky Shore	Marine and Estuarine	Intertidal	Rocky Shore	Regularly Flooded and Irregularly Flooded
Tidal Flat	Marine and Estuarine	Intertidal	Unconsolidated Shore	Regularly Flooded
	Riverine	Tidal	Unconsolidated Shore	Regularly Flooded
Salt Marsh and Brackish Marsh	Estuarine	Intertidal	Emergent Wetland (includes Scrub-Shrub Wetland)	Regularly Flooded and Irregularly Flooded
Tidal Fresh Marsh	Riverine	Tidal	Emergent Wetland	Regularly Flooded
	Palustrine		Emergent Wetland	Seasonally Flooded-Tidal
Tidal Swamp	Palustrine		Forested Wetland	Seasonally Flooded-Tidal

England and Long Island. Many of these exposed bedrock cliffs and outcroppings are vegetated by dense growths of macroalgae, especially brown algae (Phaeophycophyta). Rockweeds (*Fucus* spp. and *Ascophyllum nodosum*) are most common. Although long considered as marine intertidal habitats, rocky shores are now viewed as a type of coastal wetland.

Tidal Flats

Tidal flats occur throughout the coastal zone. They are nearly level areas largely devoid of macrophytic vegetation. In more saline areas, some flats are covered by macroscopic algae, such as sea lettuce (*Ulva lactuca*) and other green algae (*Enteromorpha* spp.), or by microscopic diatoms. Isolated

clumps of smooth cordgrass (*Spartina alterniflora*) may occasionally be found at the highest elevations. In brackish and tidal freshwater areas, numerous low-growing plants may be locally abundant on tidal flats. These plants include eastern lilaeopsis (*Lilaeopsis chinensis*), mudwort (*Limosella subulata*), pygmyweed (*Crassula aquatica*), seaside crowfoot (*Ranunculus cymbalaria*), riverbank quillwort (*Isoetes riparia*), and Parker's pipewort (*Eriocaulon parkeri*). Yet the majority of flats appear as large expanses of mud, sand, gravel, or various mixtures at low tide. At the lowest levels nearest the water, aquatic bed plants may be exposed by the lowest annual tides. These plants include eelgrass (*Zostera marina*) and widgeon grass

4. Rockweed-covered shores are a
common sight along Maine's rocky coast.

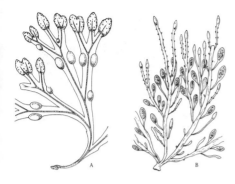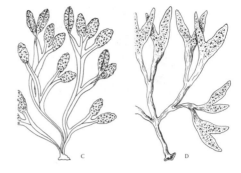

5. Common rockweeds of the Northeast: (A) *Fucus vesiculosus;* (B) *Ascophyllum* *nodosum;* (C) *Fucus spiralis;* and (D) *Fucus edentatus.*

(*Ruppia maritima*) in saline areas, widgeon grass and sago pondweed (*Potamogeton pectinatus*) in brackish waters, and tape grass or wild celery (*Vallisneria americana*) and other pondweeds (*Potamogeton* spp.) in fresh tidal waters.

Salt Marshes

Salt marshes are intertidal areas vegetated by grasses and other salt-tolerant plants, collectively called *halophytes*. These marshes become the dominant coastal wetland type from New Hampshire south, occurring as a broad band behind barrier beaches and islands and at the mouths of coastal rivers in high-salinity waters. The dominant plants are emergent, herbaceous vegetation; yet shrubs are often present and may be locally abundant.

Two vegetation zones are evident within salt marshes: (1) the low marsh and (2) the high marsh. The low marsh consists of a single plant species—smooth cordgrass (*Spartina alterniflora*)—which grows to a height of 6 feet (tall growth form). This zone is flooded at least once daily by the tides. A recent study in Connecticut found that salt hay grass (*Spartina patens*) may also occur within this zone. The high marsh is floristically more diverse with several species abundant, including the short growth form of smooth cordgrass (less than 1½ feet tall), salt hay grass, spike grass (*Distichlis spicata*), black grass (*Juncus gerardii*), and high-tide bush (*Iva frutescens*). Depressions within the high marsh called *pannes* retain salt water for longer periods than the rest of the marsh, and evaporation concentrates salts over time. Salinities here may exceed 50 parts per thousand during summer. Few plants can tolerate these conditions, but among them are the short form of smooth cordgrass,

6. Tidal flats are exposed at low tide. They are common between coastal marshes and deep water.

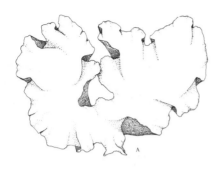

7. Common green algae of tidal flats: (A) sea lettuce (*Ulva lactuca*) and (B) *Enteromorpha intestinalis*.

glassworts (*Salicornia* spp.), plantains (*Plantago* spp.), arrow grasses (*Triglochin* spp.), seaside gerardia (*Agalinis maritima*), sea blites (*Suaeda* spp.), and sea lavenders (*Limonium* spp.). Blue-green algae (Schizophyceae) may form dense mats in these depressions. At the upland edge of salt marshes, plant diversity is greatest, with other plants common, including salt marsh bulrush (*Scirpus robustus*), groundsel tree (*Baccharis halimifolia*), seaside goldenrod (*Solidago sempervirens*), switchgrass (*Panicum virgatum*), slough grass (*Spartina pectinata*), and high-tide bush, which also grows on banks along many mosquito ditches. Where freshwater runoff from the upland is heavy or at seepage sites, brackish and freshwater plants, such as common three-square (*Scirpus pungens* = *S. americanus*), marsh fern (*Thelypteris thelypteroides*), rose

mallow (*Hibiscus moscheutos*), or narrow-leaved cattail (*Typha angustifolia*), may be found.

Geographically, some salt marsh plants attain their northern or southern range limits within the Northeast. Northern plants at their southern limit include baltic rush (*Juncus balticus*), silverweed (*Potentilla anserina*), alkali grasses (*Puccinellia* spp.), salt marsh sedge (*Carex paleacea*), seaside plantain (*Plantago maritima*), and seaside arrow grass (*Triglochin maritimum*). Southern species are represented by big cordgrass (*Spartina cynosuroides*), perennial salt marsh aster (*Aster tenuifolius*), sea lavender (*Limonium carolinianum*), marsh pinks (*Sabatia dodecandra* and *S. stellaris*), salt marsh fimbristylis (*Fimbristylis castanea*), seashore mallow (*Kosteletzkya virginica*), black needlerush (*Juncus roemerianus*), wax myrtle (*Myrica cerifera*), and others.

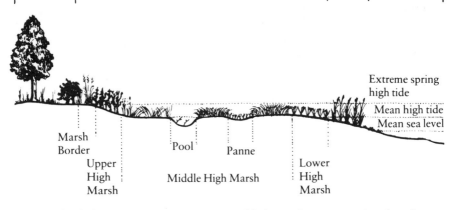

8. Generalized plant zonation in northeastern salt marshes: (1) low marsh and (2) high marsh. The high marsh can be further subdivided into several subzones. Pools and shallow depressions called "pannes" occur within the high marsh.

Notice on the accompanying chart that plant diversity increases toward upland. Within individual wetlands, high marsh communities are intermixed, forming a complex mosaic.

Brackish Marshes

Brackish marshes are found along coastal rivers immediately upstream of salt marshes, where mixing of salt water with fresh water is greater. The vegetation of these marshes is highly varied, due in large part to the broad range of salinities characteristic of this coastal wetland type. Salinities vary from moderately high (18 parts per thousand) to essentially fresh water (0.5 parts per thousand). At more seaward locations, many salt marsh plants are abundant, whereas at upstream areas where waters are fresh in winter and spring and become slightly brackish in summer, many freshwater plants are common. Consequently, the brackish marshes represent a vegetative continuum characterized by a gradual intermixing of salt marsh species with tidal fresh marsh plants. Major dominant species include salt hay grass, salt marsh bulrush, narrow-leaved cattail, common reed (*Phragmites australis*), Olney three-square (*Scirpus americanus* = *S. olneyi*), rose mallow, seaside goldenrod, and big cordgrass. From Maryland south, black needlerush becomes a major dominance type in brackish

	Plant Community	
Vegetative Zone	Dominant Plants	Common Associates
Low Marsh	smooth cordgrass (tall form)	smooth cordgrass (short form), rockweed (*Fucus vesiculosus*—locally), and other algae
High Marsh		
Lower High Marsh	smooth cordgrass (short form)	glassworts, salt hay grass, sea lavender, and filamentous green algae
Middle High Marsh	salt hay grass	spike grass (often codominant), sea lavender, black grass, marsh orach, and sea blites
Panne	glassworts, smooth cordgrass (short form), seaside plantain, and blue-green algae	seaside gerardia, sea blites, seaside arrow grass, sea lavender, spike grass, sea milkwort, and salt hay grass
Pool	widgeon grass	
Upper High Marsh	black grass	spike grass, salt hay grass, perennial salt marsh aster, sea lavender, high-tide bush, salt marsh bulrush, seaside arrow grass, seaside goldenrod, and seashore alkali grass
Marsh Border	switchgrass, slough grass, common reed, high-tide bush, and groundsel tree; in seepage areas: narrow-leaved cattail, three-squares, and rose mallow	seaside goldenrod, grass-leaved goldenrod, salt hay grass, annual marsh pink, creeping bent grass, red fescue, foxtail grass, American germander, hedge bindweed, poison ivy, marsh fern, baltic rush (New England), sweet gale (New England), northern bayberry, salt marsh fimbristylis (Long Island south), and wax myrtle (Delaware south)

marshes. Other common plants are smooth cordgrass, spike grass, creeping bent grass (*Agrostis stolonifera* var. *compacta*), water hemp (*Amaranthus cannabinus*), soft-stemmed bulrush (*Scirpus validus*), perennial salt marsh aster, alkali grasses, slough grass, spike-rushes (*Eleocharis* spp.) and water parsnip (*Sium suave*). Plants bordering the upland may include switchgrass, purple loosestrife (*Lythrum salicaria*), and others.

9. Salt marsh bordering the North River near Norwell, Massachusetts. The tall form of smooth cordgrass (*Spartina alterniflora*) dominates the river bank, while salt hay grass (*Spartina patens*) forms cowlicks in the high marsh.

10. Brackish marsh along Taylor River in
New Hampshire. Narrow-leaved cattail
(*Typha angustifolia*), smooth cordgrass
(*Spartina alterniflora*), and salt marsh
bulrush (*Scirpus robustus*) are abundant.

11. Tidal fresh marsh bordering
Crosswicks Creek, a tributary of the
Delaware River, near Trenton, New Jersey.
A distinct zonation pattern is evident, with
spatterdock (*Nuphar luteum*) dominating
the low marsh and other plants forming a
mixed community in the high marsh.

Tidal Fresh Marshes

Beyond the limit of saltwater penetration in large coastal rivers, tidally influenced freshwater wetlands may exist. Such rivers include the Kennebec River in Maine, the Merrimack and North rivers in Massachusetts, the Connecticut River in Connecticut, the Hudson River in New York, the Delaware, Great Egg Harbor, Mullica, and Maurice rivers in New Jersey, and the Choptank, Nanticoke, Patuxent, and Pocomoke rivers in Maryland. Flooding of adjacent wetlands is influenced by the tides as well as by seasonal river overflows.

Vegetation in tidal fresh marshes is extremely diverse, with many species intermixed. Along the water's edge in regularly flooded areas, wild rice (*Zizania aquatica*), big-leaved arrowhead (*Sagittaria latifolia*), spatterdock (*Nuphar luteum*), arrow arum (*Peltandra virginica*), pickerelweed (*Pontederia cordata*), water hemp, and soft-stemmed bulrush may be present. Just above this zone, other plants, such as sweet flag (*Acorus calamus*), river bulrush (*Scirpus fluviatilis*), water parsnip, sneezeweed (*Helenium autumnale*), square-stemmed monkeyflower (*Mimulus ringens*), rice cutgrass (*Leersia oryzoides*), tearthumbs (*Polygonum arifolium* and *P. sagittatum*), water smartweeds (*Polygonum punctatum* and *P.*

hydropiperoides), common three-square, and bur marigold (*Bidens laevis*), may coexist with wild rice, arrow arum, and big cordgrass. Further landward, narrow-leaved cattail, broad-leaved cattail (*Typha latifolia*), marsh fern, tussock sedge (*Carex stricta*), royal fern (*Osmunda regalis*), purple loosestrife, and bluejoint (*Calamagrostis canadensis*) become abundant. Many other emergent plants coexist with these more common species. Shrubs, such as swamp rose (*Rosa palustris*), buttonbush (*Cephalanthus occidentalis*), willows (*Salix* spp.), and silky dogwood (*Cornus amomum*), and saplings of red maple (*Acer rubrum*) may be scattered in clusters at higher elevations within these marshes or occur along the upland edges. In Delaware, wax myrtle forms a dense shrub thicket along the landward edge of many tidal fresh marshes. Many tidal fresh marshes grade naturally into tidal swamps.

Tidal Swamps

Forested wetlands along freshwater stretches of coastal rivers may also be under tidal influence. These wetlands are flooded for shorter periods than the lower-lying tidal marshes. Vegetation is similar to that of nontidal swamps and may include overstory species of red maple, green ash (*Fraxinus pennsylvanica* var. *subintegerrima*), and black gum (*Nyssa sylvatica*);

12. Tidal swamps are especially common
in the Chesapeake Bay area. Here a tidal
swamp lies behind a regularly flooded
spatterdock (*Nuphar luteum*) marsh
bordering the Nanticoke River near
Seaford, Delaware.

understory shrubs of southern arrowwood (*Viburnum dentatum*), alders (*Alnus rugosa* and *A. serrulata*), sweet pepperbush (*Clethra alnifolia*), spicebush (*Lindera benzoin*), common winterberry (*Ilex verticillata*), and swamp azalea (*Rhododendron viscosum*); and emergent plants, such as tussock sedge, skunk cabbage (*Symplocarpus foetidus*), jewelweed (*Impatiens capensis*), and cardinal flower (*Lobelia cardinalis*).

Coastal Aquatic Beds

Aquatic beds occur in pools and ponds within tidal marshes and in adjacent coastal waters. They are formed by colonies of floating-leaved and/or submerged (underwater) plants. Composition of these beds varies considerably from the most saline waters nearest the ocean to the tidal fresh riverine waters upstream. Aquatic beds in the more saline waters are dominated by two vascular plants— eelgrass (*Zostera marina*) and widgeon grass (*Ruppia maritima*). As salinity decreases upstream, widgeon grass remains an important brackish species and is joined by several other dominant plants, including sago pondweed (*Potamogeton pectinatus*), clasping-leaved pondweed or redhead-grass (*Potamogeton perfoliatus*), curly pondweed (*Potamogeton crispus*), horned pondweed (*Zannichellia palustris*),

and naiads (*Najas* spp.). Wild celery (*Vallisneria americana*), waterweeds (*Elodea* spp.), and coontail (*Ceratophyllum demersum*) are also present in slightly brackish waters, but they are more abundant in tidal fresh waters. Other freshwater aquatic bed plants include clasping-leaved pondweed, several other pondweeds (*Potamogeton epihydrus, P. pusillus, P. foliosus*, and others), bullhead lily (*Nuphar luteum* ssp. *variegatum*), and white water lily (*Nymphaea odorata*). A few introduced (nonnative) aquatic plants have established local populations in some coastal waterbodies, but they are uncommon throughout the Northeast. For example, hydrilla (*Hydrilla verticillata*) is locally common in the Potomac River in Maryland, and water chestnut (*Trapa natans*) is abundant along the Hudson River in New York. Hydrilla is a native of Southeast Asia, and water chestnut comes from Eurasia.

Identification of
Coastal Wetland Plants

To identify tidal wetland plants common in the Northeast, seven keys are provided. Because plants flower at different times during the growing season and flowers may not be present at the time of observation, vegetative characteristics are emphasized in most of the keys. When flowers are present, many plants can be more easily recognized, with little consideration of other characteristics. Consequently, one key (Key G) is based largely on flower characteristics. This flower key does not include submerged plants, ferns, grasses, sedges, and similar plants but focuses on marsh plants with conspicuous flowers. In all of the keys, the use of technical terms has been minimized. A glossary defining technical terms is provided near the end of this book, and a scale to aid in measurements is included on the inside of the back cover.

How to Use the Keys

A general key to the seven keys is presented to guide you to the appropriate plant identification key. This "Key to Subsidiary Keys" separates plants first on the basis of habitat (i.e., whether growing in permanent open water or in periodically flooded areas), then on life form (e.g., woody or herbaceous plants) or on flowers, if present, and so forth. Each key consists of a series of couplets, for example, 1–1, 2–2, 3–3, and so on. Each couplet contains contrasting statements about the plant. After consulting the Key to Subsidiary Keys, you should start at the first couplet in the appropriate plant identification key (Keys A–G) and begin matching the couplet statements to the plant in hand. As you work down through the key, you will eventually reach a couplet where a specific plant is referenced. You should then turn to the designated page where the plant is illustrated and better described. In the plant descriptions, you will notice the "Similar species" subheading. It is important to read this section to distinguish the illustrated and described plant from related species or other plants resembling it. If the plant you have does not meet the description in the text or resemble the illustration, you need to go through the key again, carefully checking characteristics.

The provided keys are intended for identifying aquatic vascular plants, herbaceous plants, and shrubs that are common in tidal marshes and adjacent coastal waters. Trees found in and around tidal fresh marshes are not generally included. Only tree species that also occur as shrubs in these marshes are represented. Other plants strictly associated with tidal swamps or uncommon in tidal fresh marshes may be identified using *Freshwater Wetlands: A Guide to Common Indicator Plants of the Northeast* (Magee 1981) or other field guides.

Overview of Plant Characteristics

Before using the keys, it may be useful to review some of the more important features of wetland plants discussed and illustrated below.

Life Form

Life form relates to the growth form of a plant. Five life forms are generally recognized in tidal wetlands: (1) aquatic plants, (2) emergent plants, (3) shrubs, (4) trees, and (5) vines. Aquatic plants grow in permanently flooded waters, either free-floating, submerged (underwater), or with floating leaves at the surface and stems rooted to the underlying substrate. Emergent plants are erect, herbaceous (nonwoody) plants that have all or part of their stems and leaves standing above the water's surface or grow on the surface of intertidal areas. Shrubs are woody plants less than 20 feet in height, usually with multiple stems, but also including saplings of tree species. Trees are woody plants 20 feet or greater in height, having a single main stem (trunk). Vines are woody or herbaceous plants that climb or twine around the stems of other plants.

Leaf Types

Leaves may be conspicuous as in most wetland plants or reduced to spines or scales as in saltwort and glassworts, respectively. Simple leaves are leaves that are not divided into more than one part, whereas compound leaves are divided into two or more distinct and separate parts called leaflets. Lobed leaves are simple leaves that have shallow or deep indentations forming lobes but are not divided into separate parts. Entire leaves are leaves with smooth margins, unbroken by indentations. Toothed leaves have margins indented by fine or coarse teeth or have scalloped edges. Leaves also take on a variety of shapes, including threadlike, grasslike, linear, lance-shaped, egg-shaped, spoon-shaped, heart-shaped, arrowhead-shaped, and sword-shaped. Fleshy-leaved plants have succulent leaves that are fleshy in texture. Leaves are attached to the stem in different ways. Sessile leaves are leaves that are directly joined to the stem, without a stalk (petiole). Petioled leaves are connected to the stem by a stalk (petiole).

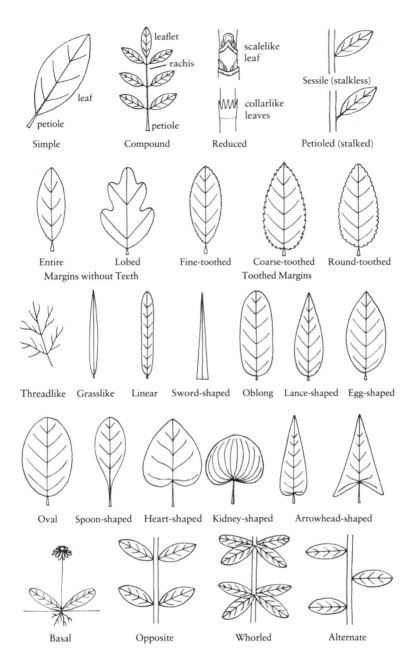

Leaf types

leaflet
rachis
leaf
petiole
petiole

scalelike leaf
collarlike leaves

Sessile (stalkless)
Petioled (stalked)

Simple   Compound   Reduced   Petioled (stalked)

Entire   Lobed   Fine-toothed   Coarse-toothed   Round-toothed
Margins without Teeth   Toothed Margins

Threadlike   Grasslike   Linear   Sword-shaped   Oblong   Lance-shaped   Egg-shaped

Oval   Spoon-shaped   Heart-shaped   Kidney-shaped   Arrowhead-shaped

Basal   Opposite   Whorled   Alternate

13. Leaf types and arrangements.

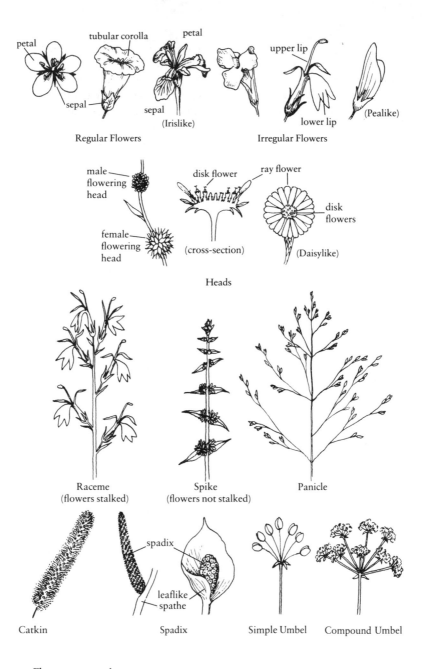

petal
sepal
tubular corolla
petal
(Irislike)
sepal
upper lip
lower lip
(Pealike)

Regular Flowers

Irregular Flowers

male flowering head
female flowering head
disk flower
ray flower
disk flowers
(cross-section)
(Daisylike)

Heads

Raceme
(flowers stalked)

Spike
(flowers not stalked)

Panicle

Catkin

spadix
leaflike spathe

Spadix

Simple Umbel

Compound Umbel

14. Flower types and arrangements.

Leaf Arrangements

Leaves are arranged on plants in four basic ways: (1) basal, (2) opposite, (3) whorled, and (4) alternate. Basal leaves grow directly from the roots of the plant; they do not arise from the stem. When leaves grow in pairs across from each other along a stem, the leaves are oppositely arranged. If the leaves grow in clusters of three or more around the stem, they are whorled. When leaves grow singly on the stem and vary in position from one side to the next up the stem, they are alternately arranged.

Flower Types and Arrangements

At first glance, flowers are either regularly or irregularly shaped. Regular flowers are radially symmetrical with distinct petals or petallike parts surrounding the center of the flower. Each petal or petallike part is alike in shape, size, and color. Irregular flowers are not radially symmetrical, and their petals or petallike parts are not alike but differ in shape, size, and/or color. Tubular flowers have petals fused into lobes that are joined at the base forming a tube. Some tubular flowers have distinct upper and lower lobes called lips. Inconspicuous flowers do not have petals, or their petals or petallike parts are so small that they cannot be readily observed.

Flowers are arranged in several ways—many grow singly along the stem, whereas others occur in clusters of various types. The latter include heads, panicles, racemes, spikes, and umbels among others. A head is a rounded or flat-topped cluster of sessile flowers, as found in many asters. A panicle is a branched inflorescence with a central axis, as present in many grasses. A raceme is an unbranched elongated inflorescence with lateral flowers. A spike is a type of raceme with sessile flowers. Some plants possess a single terminal spike, and others have numerous small spikes called spikelets branching from a central axis or side branches. An umbel is an inflorescence with several branches arising from the end of a peduncle (flowering stalk), as found in water parsnip.

Distinguishing among Grasses, Sedges, and Rushes

Many people have difficulty separating grasses, sedges, and rushes from one another. It is, therefore, useful to review the general differences between them. *Grasses.* Grasses have jointed, hollow stems that are round in cross-section. Leaves are distinctly two-ranked and connected to the stem by open sheaths. Leaves possess a ligule (membranous or hairy appendage) at the junction of the leaf and leaf sheath. Flowers are born in spikelets, and each flower consists of two glumes (bracts) and one or more florets (each of which has two different bracts, the lemma and the palea, and a flower). Fruits are grainlike seeds.

Sedges. Sedges have solid, triangular stems for the most part. Leaves are three-ranked with closed sheaths. Flowers are born in the axils of overlapping scales, which may collectively resemble a bud. Each flower consists of a single pistil with two to three stigmas and one to three stamens. Fruits are lens-shaped or three-angled nutlets called *achenes.*

Rushes. Rushes have solid stems, mostly round in cross-section. No ligule is present. They have regular flowers with three sepals, three petals, three or six stamens, and a fruit capsule. Fruit capsules are usually three-valved but sometimes one-celled. Rushes can be easily separated from grasses and sedges by their multiseeded capsules, because grasses and sedges have only one seed per flowering scale.

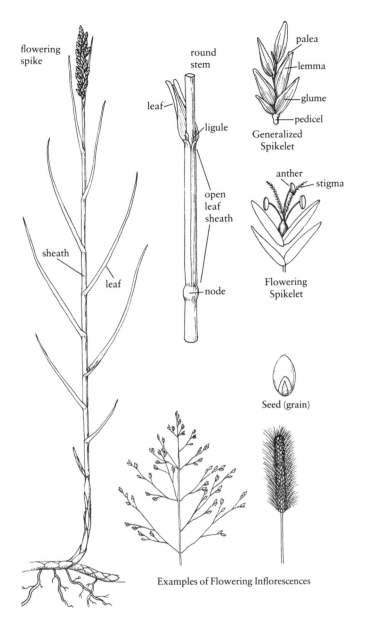

flowering
spike

round
stem

leaf

ligule

open
leaf
sheath

node

sheath

leaf

palea

lemma

glume

pedicel

Generalized
Spikelet

anther

stigma

Flowering
Spikelet

Seed (grain)

Examples of Flowering Inflorescences

15. Characteristics of grasses.

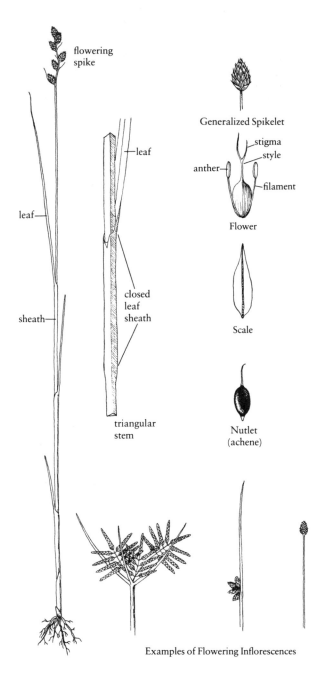

flowering
spike

leaf

leaf

sheath

leaf

closed
leaf
sheath

triangular
stem

Generalized Spikelet

stigma
style
anther
filament

Flower

Scale

Nutlet
(achene)

Examples of Flowering Inflorescences

16. Characteristics of sedges.

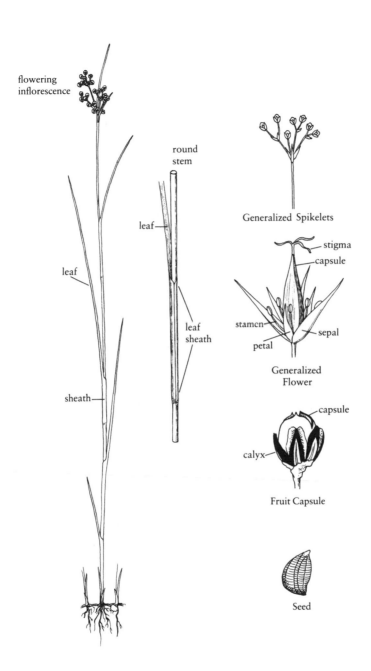

flowering
inflorescence

round
stem

leaf

leaf

sheath

leaf
sheath

Generalized Spikelets

stigma
capsule

stamen
petal
sepal

Generalized
Flower

capsule

calyx

Fruit Capsule

Seed

17. Characteristics of rushes.

Artificial Keys to Coastal Wetland Plants of the Northeastern United States

Key to Subsidiary Keys*

1. Plants growing in permanent open water   Key A
1. Plants not growing in permanent open water but characteristic of alternately flooded and exposed areas
   2. Grasses, sedges, rushes, ferns, and other plants with indistinct flowers or plants not in flower
      3. Plants woody or with woody-like stems   Key B
      3. Plants herbaceous
         4. Leaves and/or stems fleshy   Key C
         4. Leaves and stems not fleshy
            5. Plants growing as climbing vines   Key D
            5. Plants growing erect and free-standing
               6. Low and medium-height emergent plants ranging to 5 feet tall at maturity   Key E
               6. Tall emergent plants, exceeding 5 feet at maturity   Key F
   2. Marsh plants in flower, excluding horsetails, quillworts, grasses, sedges, rushes, and ferns   Key G

Key A Plants of Permanent Coastal Waters

1. Plant free-floating, not attached to substrate
   2. Plant composed of leaflike thallus, without stem   Big Duckweed, *Spirodela polyrhiza* (p. 93)

   2. Plant with much branched stem and compound, linear, toothed leaves   Coontail, *Ceratophyllum demersum* (p. 95)

1. Plant not free-floating but attached to substrate
   3. Leaves all or mostly submerged; those not submerged floating at or near the water's surface
      4. Linear basal leaves only

*Note: If plant is a tree, refer to *Freshwater Wetlands: A Guide to Common Indicator Plants of the Northeast* (Magee 1981).

5. Leaves short (1–2 inches) and thin (½–1 inch wide) in the middle

Parker's Pipewort, *Eriocaulon parkeri* (p. 186)

5. Leaves very long (up to 6 feet) and thin (⅕–⅓ inch wide)

Wild Celery, *Vallisneria americana* (p. 92)

4. Leaves not basal only but variously arranged along stem
    6. Narrow leaves arranged in whorls of threes

Waterweed, *Elodea canadensis* (p. 91)

    6. Leaves not in whorls of threes
        7. Narrow leaves oppositely arranged
            8. Leaves 1¼–4 inches long, threadlike (very narrow)

Horned Pondweed, *Zannichellia palustris* (p. 88)

            8. Leaves less than 1¼ inches long

Naiad, *Najas guadalupensis* (p. 90)

        7. Leaves alternately arranged
            9. Flowers and fruits enclosed by a leaf sheath; leaves flattened and linear, up to 2 feet long; common aquatic plant of saline waters

Eelgrass, *Zostera marina* (p. 89)

            9. Flowers and fruits not enclosed in a leaf sheath; leaves egg-shaped or threadlike
                10. Flowers and fruits on a stalk; leaves threadlike

Widgeon Grass, *Ruppia maritima* (p. 87)

                10. Flowers and fruits in spikes or heads
                    11. Both submerged and floating leaves present

Ribbonleaf Pondweed, *Potamogeton epihydrus* (p. 84)

                    11. Submerged leaves only; floating leaves absent
                        12. Leaf margins wavy or curly

Curly Pondweed, *Potamogeton crispus* (p. 83)

                        12. Leaf margins not wavy or curly but smooth
                            13. Submerged leaves egg-shaped or rounded, generally wider than ⅕ inch

Clasping-leaved Pondweed, *Potamogeton perfoliatus* (p. 86)

34

13. Submerged leaves threadlike, less than
    ⅕ inch wide
                                           Sago Pondweed,
                                           Potamogeton
                                           pectinatus (p. 85)

3. Leaves not all or mostly submerged but floating on water's
   surface or standing erect out of shallow water
   14. Leaves mostly floating on surface of water
       15. Leaves roundish, often purplish below; large,
           fragrant, showy white flower   White Water Lily,
                                                  Nymphaea odorata
                                                  (p. 94)

       15. Leaves somewhat heart-shaped; large yellow
           flower with five to six petals   Bullhead Lily, *Nuphar*
                                                   luteum ssp.
                                                   variegatum (p. 202)

   14. Leaves mostly emergent, standing erect out of shallow
       water
       16. Linear leaves not basal, triangular in cross-section
           and clasping the stem   Great Bur-reed,
                                                   Sparganium
                                                   eurycarpum (p. 166)

       16. Leaves basal or no apparent leaves
           17. No apparent leaves; stem with terminal,
               budlike, flowering spike   Spike-rushes,
                                                   Eleocharis spp.
                                                   (pp. 116–17)

           17. Basal leaves only
               18. Elongate, linear, basal leaves (up to 6 feet)
                   forming long sheaths along stem and
                   branching outward in an alternately
                   arranged fashion; numerous minute
                   flowers borne terminally on fertile stalk;
                   brackish and tidal fresh marsh plant   Narrow-leaved Cattail,
                                                            Typha angustifolia
                                                            (p. 99)

               18. Leaves not elongate
                   19. Linear, basal leaves (up to 2 feet long)
                       and flattened (to ⅗ inch wide); three-
                       petaled white flowers arranged in whorls
                       of one to three on stalk arising between
                       basal leaves; tidal fresh marsh plants   Arrowheads,
                                                                 Sagittaria spp.
                                                                 (p. 168)

19. Leaves not linear
  20. Leaves egg-shaped
    21. Leaves rolled inwardly where attached to long fleshy stalk; minute yellow flowers borne on a fleshy stalk

                                   Golden Club, *Orontium aquaticum* (p. 182)

    21. Leaves not so; three-petaled white flowers arranged in whorls on fertile stalk

                                   Arrowheads, *Sagittaria* spp. (p. 168)

  20. Leaves not egg-shaped
    22. Leaves heart-shaped and large
      23. Leaves with midrib; single, large yellow flower borne on a stalk arising from substrate

                                   Spatterdock, *Nuphar luteum* (p. 202)

      23. Leaves without midrib; numerous small purplish flowers borne on a terminal spike

                                   Pickerelweed, *Pontederia cordata* (p. 188)

    22. Leaves arrowhead-shaped and large
      24. Leaves distinctly three-nerved, with several smaller veins and another large, thick vein along the leaf margin; leaves somewhat leathery or waxy; small and inconspicuous flowers borne in a green spathe

                                   Arrow Arum, *Peltandra virginica* (p. 183)

      24. Leaves not three-nerved, but all veins same size, narrow, and threadlike; leaves not waxy or leathery; white three-petaled flowers arranged in whorls of two to fifteen on a fertile stalk arising from substrate

                                   Big-leaved Arrowhead, *Sagittaria latifolia* (p. 168)

Key B   Woody Plants and Herbaceous Plants with Woody-like Stems

1. True woody shrubs or saplings of trees (less than 20 feet in height)
  2. Branches with thorns

                                   Swamp Rose, *Rosa palustris* (p. 205)

2. Branches without thorns
   3. Leaves simple
      4. Leaves leathery in texture and/or aromatic when crushed
         5. Leaves deciduous; fruit nutlet; plant occurring from Long Island north — Sweet Gale, *Myrica gale* (p. 125)

         5. Leaves evergreen and leathery; fruit waxy and ball-like; plant occurring from southern New Jersey south — Wax Myrtle, *Myrica cerifera* (p. 195)

      4. Leaves not leathery or aromatic
         6. Leaves oppositely arranged
            7. Leaves divided into three to five shallow, pointed lobes; tree sapling of tidal fresh marshes — Red Maple, *Acer rubrum* (p. 211)

            7. Leaves not divided into three to five lobes
               8. Leaf margins coarse-toothed
                  9. Leaves thick and somewhat fleshy; common shrub of salt and brackish marshes — High-tide Bush, *Iva frutescens* (p. 156)

                  9. Leaves thin and not fleshy; shrub of tidal fresh marshes and swamps — Northern Arrowwood, *Viburnum recognitum* (p. 233)

               8. Leaf margins smooth
                  10. Leaf veins, usually three to five pairs, strongly following leaf margins; bark gray with reddish purple stripes, twigs dull reddish purple; shrub of upper tidal fresh marshes — Silky Dogwood, *Cornus amomum* (p. 219)

                  10. Leaf veins not strongly following leaf margins; petioles often red; old bark grayish brown and flaky, twigs grayish brown to purplish with elongate light dots (lenticels); shrub typical of lower tidal fresh marshes — Buttonbush, *Cephalanthus occidentalis* (p. 230)

6. Leaves alternately arranged
    11. Leaves toothed mostly above middle with wedge-shaped bases
        12. Leaves thick and few deep-toothed; plant of salt, brackish, and tidal fresh marshes ... Groundsel Tree, *Baccharis halimifolia* (p. 153)

        12. Leaves thin and fine-toothed; plant of tidal fresh marshes and swamps ... Sweet Pepperbush, *Clethra alnifolia* (p. 220)

    11. Leaves toothed along entire margin
        13. Somewhat heart-shaped leafy structures (stipules) present at leaf bases; leaves fine-toothed, narrowly lance-shaped, tapering to a long point ... Black Willow, *Salix nigra* (p. 194)

        13. Leafy stipules absent and leaves not narrowly lance-shaped and long-pointed
            14. Leaves tapering distally to a distinct short point, with somewhat wedge-shaped bases; bark smooth and gray; fruit red berry ... Common Winterberry, *Ilex verticillata* (p. 210)

            14. Leaves tapered or somewhat rounded, lacking the distinct short point; bark dark gray marked with light dots (lenticels); small cones usually present; fruit nutlet ... Smooth Alder, *Alnus serrulata* (p. 196)

3. Leaves not simple but compound—divided into separate leaflets
    15. Leaves alternately arranged
        16. Three leaflets ... Poison Ivy, *Toxicodendron radicans* (p. 209)

        16. Eleven to thirty-five leaflets ... False Indigo, *Amorpha fruticosa* (p. 207)

    15. Leaves oppositely arranged

17. Five to nine leaflets, usually seven; twigs gray and
smooth; fruit winged seed (samara); plant a tree
sapling — Green Ash, *Fraxinus pennsylvanica* var. *subintegerrima* (p. 221)

17. Five to eleven leaflets, usually seven; twigs light
brown with numerous large bumps (lenticels); soft
white pith; fruit dark purplish berry; true shrub
with multiple stems — Common Elder, *Sambucus canadensis* (p. 232)

1. Flowering herbaceous plants with woody-like stems; not
shrubby in appearance but an erect emergent plant
  18. Leaves oppositely arranged or sometimes in whorls of
  threes or fours
    19. Stems slender and arching; leaves lance-shaped
    and short-stalked — Water Willow, *Decodon verticillatus* (p. 215)

    19. Stems erect; leaves with rounded or heart-shaped
    bases, somewhat clasping stem — Purple Loosestrife, *Lythrum salicaria* (p. 216)

  18. Leaves alternately arranged, usually three-lobed and
  toothed
    20. Leaves smooth above and fine hairy below; upper
    leaves may be egg-shaped without lobes — Rose Mallow, *Hibiscus moscheutos* (p. 138)

    20. Leaves hairy on both sides
      21. Stem and leaves velvety smooth hairy — Marsh Mallow, *Althaea officinalis* (p. 137)

      21. Stem and leaves hairy and rough, not velvety
      smooth — Seashore Mallow, *Kosteletzkya virginica* (p. 139)

Key C   Fleshy Plants

1. Plants with no apparent leaves; leaves reduced to sharp
fleshy spines or minute scales

2. Leaves reduced to sharp fleshy spines	Saltwort, *Salsola kali* (p. 130)
2. Leaves reduced to minute scales; stem jointed	
3. Perennial plant with solitary fleshy stem with a woody core; stem roots at nodes forming a mat	Perennial Glasswort, *Salicornia virginica* (p. 129)
3. Annual plant with branching fleshy stem	
4. Stem greater than ⅕ inch wide; scales sharp-tipped	Bigelow's Glasswort, *Salicornia bigelovii* (p. 128)
4. Stem less than ⅕ inch wide; scales round-tipped	Common Glasswort, *Salicornia europaea* (p. 128)
1. Plants with normal leaves, not reduced	
5. Linear leaves	
6. Leaves oppositely arranged and sessile	
7. Leaves with triangle-shaped structures (stipules) at leaf bases	Salt Marsh Sand Spurrey, *Spergularia marina* (p. 133)
7. Leaves without triangular stipules at leaf bases	
8. Plant of freshwater tidal mud flats; stem usually less than 4 inches tall	Pygmyweed, *Crassula aquatica* (p. 204)
8. Plant of salt and brackish marshes	
9. Leaves with blunt or round tips, up to ⅘ inch long and ¼ inch wide; perennial	Sea Milkwort, *Glaux maritima* (p. 142)
9. Leaves with tapered tips, up to 1⅖ inches long and to 1/16 inch wide, sometimes alternately arranged on end of branches; annual	Seaside Gerardia, *Agalinis maritima* (p. 148)
6. Leaves not oppositely arranged	
10. Basal leaves only, sometimes forming long sheaths along stem and branching outward in an alternately arranged fashion but all leaves arising from and attached to base of plant	
11. Leaves sword-shaped	Blue Flag, *Iris versicolor* (p. 192)

11. Leaves not sword-shaped
    12. Leaves flattened and up to 6 feet long,
        extending along stem as long sheaths
        before branching outward in an
        alternately arranged fashion   Cattails, *Typha* spp.
                                               (p. 99)

    12. Leaves not flattened
        13. Leaves hollow, divided into four
            cavities (in cross-section), sharp-
            pointed, and pale green; plant of
            tidal shores in fresh water   Riverbank Quillwort,
                                           Isoetes riparia (p. 162)

        13. Leaves not so; salt and brackish
            marsh plant
            14. Leaves with conspicuous
                sheaths, somewhat triangular
                in cross-section, and up to 20
                inches long   Seaside Arrow Grass,
                                           Triglochin maritimum
                                         (p. 100)

            14. Leaves without sheaths,
                triangular in cross-section, and
                up to 6 inches long   Seaside Plantain,
                                           Plantago maritima
                                         (p. 150)

10. Leaves not basal but alternately arranged or in whorls
   along stem
    15. Leaves arranged in whorls and sessile
        16. Leaves arranged singly in whorls around the stem;
            stem creeping or erect; leaves sharp-tipped   Sea Blite, *Suaeda*
                                           linearis (p. 131)

        16. Leaves arranged in whorls of twos or more; leaf
            nodes swollen and somewhat cup-shaped around
            leaf bases   Salt Marsh Sand
                                         Spurrey, *Spergularia*
                                         marina (p. 133)

    15. Leaves alternately arranged and sessile
        17. Leaves narrowly linear; fibrous-rooted perennial
            with creeping rhizomes   Perennial Salt Marsh
                                         Aster, *Aster*
                                         tenuifolius (p. 152)

        17. Leaves narrowly linear to narrowly oblong; annual
            with short taproot   Annual Salt Marsh
                                         Aster, *Aster subulatus*
                                         (p. 151)

5. Leaves not linear
    18. Leaves spatulate (spoon-shaped) with wavy margins — Sea Rocket, *Cakile edentula* (p. 135)

    18. Leaves not spatulate
        19. Leaves arrowhead-shaped
            20. Basal leaves only, conspicuously three-nerved and 4–12 inches long; slightly brackish and tidal fresh marsh plant — Arrow Arum, *Peltandra virginica* (p. 183)

            20. Leaves not basal but arranged along erect or creeping stem and up to 3 inches long; salt and brackish marsh plant — Marsh Orach, *Atriplex patula* (p. 127)

        19. Leaves not arrowhead-shaped
            21. Leaves heart-shaped
                22. Leaves with prominent midrib — Spatterdock, *Nuphar luteum* (p. 202)

                22. Leaves without prominent midrib
                    23. Leaves 2 inches or less wide and 1⅕ inches long, often kidney-shaped — Mud Plantain, *Heteranthera reniformis* (p. 187)

                    23. Leaves larger, up to 7¼ inches long, occasionally somewhat lance-shaped — Pickerelweed, *Pontederia cordata* (p. 188)

            21. Leaves not heart-shaped
                24. Leaves kidney-shaped — Mud Plantain, *Heteranthera reniformis* (p. 187)

                24. Leaves not kidney-shaped
                    25. Leaves egg-shaped
                        26. Leaves basal, smooth-margined, and thick — Golden Club, *Orontium aquaticum* (p. 182)

                        26. Leaves alternately arranged, toothed, and thin; stem somewhat transparent — Jewelweed, *Impatiens capensis* (p. 212)

                    25. Leaves not egg-shaped

27. Leaves oppositely arranged, narrowly oblong to linear, and less than ⅕ inch long — Sea Milkwort, *Glaux maritima* (p. 142)

27. Leaves alternately arranged, oblong, and 4–16 inches long — Seaside Goldenrod *Solidago sempervirens* (p. 158)

Key D   Vines and Other Climbing Plants

1. Plant a twining vine
    2. Leaves reduced to scales; stems yellow or orange; plant parasitic — Common Dodder, *Cuscuta gronovii* (p. 224)

    2. Leaves not reduced but normal, with petioles
        3. Leaves compound, divided into five to seven lance-shaped leaflets with somewhat rounded bases — Ground-nut, *Apios americana* (p. 208)

        3. Leaves simple, not compoundly divided
            4. Leaves oppositely arranged; leaf bases rounded or heart-shaped, tapering above to a slender point — Climbing Hempweed, *Mikania scandens* (p. 241)

            4. Leaves alternately arranged or in whorls of threes
                5. Leaves heart-shaped, tapering to a slender, curved point; lower leaves often in whorls of threes — Wild Yam, *Dioscorea villosa* (p. 191)

                5. Leaves arrowhead-shaped or triangular, often with somewhat squarish or heart-shaped basal lobes — Hedge Bindweed, *Calystegia sepium* (p. 223)

1. Plant not a twining vine
    6. Plants creeping or reclining on other plants when old; when young, plants erect; stem covered with weak prickles or sharp, fine spines
        7. Stem covered with weak prickles and leaves arranged in whorls of fives or sixes — Dye Bedstraw, *Galium tinctorium* (p. 231)

7. Stem covered with sharp, fine spines and leaves alternately arranged
    8. Leaves broadly arrowhead-shaped with triangular basal lobes

                                              Halberd-leaved Tearthumb, *Polygonum arifolium* (p. 198)

    8. Leaves lance-shaped with heart-shaped bases and tapering to a point above

                                              Arrow-leaved Tearthumb, *Polygonum sagittatum* (p. 200)

6. Trailing vine with petioled, arrowhead-shaped leaves and smooth stem

                                              Hedge Bindweed, *Calystegia sepium* (p. 223)

Key E   Low and Medium-height Herbaceous Plants

1. Plants less than 6 inches tall
    2. Leaves seemingly lacking but actually reduced to sheaths; plant represented by an elongate stem bearing terminal budlike spikelet covered by brown or green scales

                                              Dwarf Spike-rush, *Eleocharis parvula* (p. 116)

    2. Leaves present, either basal or along stem
        3. Basal leaves only or mostly
            4. Compound leaves divided into seven or more leaflets, silvery hairy below

                                              Silverweed, *Potentilla anserina* (p. 136)

            4. Simple leaves, not divided into leaflets
                5. Linear leaves
                    6. Leaves with four to six transverse septa (partitions) and round-tipped; plant with creeping rhizomatous stems

                                              Eastern Lilaeopsis, *Lilaeopsis chinensis* (p. 140)

                  6. Leaves without septa
                    7. Leaves blunt-tipped, growing in dense tufts

                                              Mudwort, *Limosella subulata* (p. 149)

7. Leaves not blunt-tipped but tapering
to a fine point
       Parker's Pipewort,
       Eriocaulon parkeri
       (p. 186)

5. Oval heart-shaped to kidney-shaped leaves
with round teeth; some leaves not basal
       Seaside Crowfoot,
       Ranunculus
       cymbalaria (p. 134)

3. Leaves not basal, oppositely arranged
  8. Leaves fleshy and joined at base along stem
       Pygmyweed, *Crassula*
       aquatica (p. 204)

  8. Leaves not fleshy or joined at stem
    9. Leaves entire and round-tipped
       American Waterwort,
       Elatine americana
       (p. 214)

    9. Leaves coarse shallow-toothed with
tapered tip
       Overlooked Hedge
       Hyssop, *Gratiola*
       neglecta (p. 228)

1. Plants taller than 6 inches but less than 5 feet
  10. Stems covered with delicate, sharp spines, weak prickles,
or very rough hairs
    11. Stems covered by rough hairs; leaf margins very
rough; plants erect
       Rice Cutgrass, *Leersia*
       oryzoides (p. 172)

    11. Stems covered by weak prickles or sharp spines;
plants somewhat low-growing or trailing
      12. Stems covered by weak prickles
       Dye Bedstraw, *Galium*
       tinctorium (p. 231)

      12. Stems covered by sharp spines
        13. Leaves broadly arrowhead-shaped with
triangular basal lobes
       Halberd-leaved
       Tearthumb,
       Polygonum arifolium
       (p. 198)

        13. Leaves lance-shaped with heart-shaped
bases and tapering to a point
       Arrow-leaved
       Tearthumb,
       Polygonum sagittatum
       (p. 200)

  10. Stems not covered with prickles, spines, or very rough
hairs
    14. Leaves absent or reduced and inconspicuous
      15. Stems linear (unbranched) and triangular in
cross-section

16. Stems deeply concave in cross-section	Olney Three-square, *Scirpus americanus* (p. 119)
16. Stem not deeply concave but more regularly triangular in cross-section	Common Three-square, *Scirpus pungens* (p. 120)

15. Stems linear and round in cross-section
    17. Leaves reduced to toothed, collarlike sheaths along jointed, hollow, finely ridged stems; whorls of hollow branches often present; fruiting spores borne in terminal conelike spike   Water Horsetail, *Equisetum fluviatile* (p. 161)

    17. Leaves not reduced to toothed sheaths and stems not jointed
        18. Inconspicuous flowers borne in a single terminal budlike spikelet covered by green or brown scales (linear basal leaves may be present)   Spike-rushes, *Eleocharis* spp. (p. 117)

        18. Inconspicuous flowers borne not in a single terminal spikelet but in clusters
            19. Flowers borne in elongate, drooping, open cluster of many spikelets, almost at tip of stem   Soft-stemmed Bulrush, *Scirpus validus* (p. 122)

            19. Flowers borne not in drooping clusters near tip of stem but in clusters arising from upper half of stem
                20. Plant tussock-forming   Soft Rush, *Juncus effusus* (p. 190)

                20. Plant not tussock-forming but growing from creeping rhizome   Baltic Rush, *Juncus balticus* (p. 123)

14. Leaves present and conspicuous
    21. Basal leaves only, sometimes forming long sheaths along stem but all arising from and attached to base of plant
        22. Leaves linear and elongate
            23. Leaves aromatic; midvein present but somewhat off-center   Sweet Flag, *Acorus calamus* (p. 181)

            23. Leaves not aromatic when crushed

24. Leaves narrow and up to 6 feet long, extending
along stem as long sheaths before branching
outward   Cattails, *Typha* spp.
(p. 99)

24. Leaves sword-shaped and in dense clumps   Blue Flag, *Iris
versicolor* (p. 192)

22. Leaves not linear
  25. Leaves egg-shaped
    26. Leaf bases somewhat heart-shaped   Northern Water
Plantain, *Alisma
plantago-aquatica*
(p. 167)

    26. Leaf bases tapered and inwardly rolled where
attached to stalk   Golden Club,
Orontium aquaticum
(p. 182)

  25. Leaves not egg-shaped
    27. Leaves arrowhead-shaped or triangular
      28. Leaves with four distinct, large, thick veins
(one along leaf margin) and several smaller
veins; leaves somewhat leathery or waxy in
texture   Arrow Arum,
Peltandra virginica
(p. 183)

      28. Leaves without thick veins; all veins similar
—narrow, threadlike; leaves not waxy or
leathery   Big-leaved Arrowhead,
Sagittaria latifolia
(p. 168)

    27. Leaves oval or heart-shaped, rounded at bases
      29. Leaves foul-smelling (skunklike odor)   Skunk Cabbage,
Symplocarpus foetidus
(p. 184)

      29. Leaves not foul-smelling
        30. Leaves with parallel veins; petiole
connected to end of leaf and not
continued as midrib on underside
of leaf   Pickerelweed,
Pontederia cordata
(p. 188)

        30. Leaves with veins branching near
margins; petiole extending along
underside of leaf to form a prominent
midrib   Spatterdock, *Nuphar
luteum* (p. 202)

21. Leaves not only basal but borne on and arranged along a
    stem
    31. Leaves slightly triangular in cross-section — Great Bur-reed, *Sparganium eurycarpum* (p. 166)

    31. Leaves not slightly triangular in cross-section
        32. Leaves few and round in cross-section, at least near
            leaf base; stems linear or only weakly branched
            and round in cross-section
            33. Leaves with partitions (transverse septa) at
                regular intervals — Canada Rush, *Juncus canadensis* (p. 189)

            33. Leaves without partitions
                34. Stems 16–60 inches tall, round (in cross-
                    section) and stiff, typically ending in a
                    sharp-pointed tip; brackish and salt marsh
                    plant from southeastern Delaware and
                    Maryland south — Black Needlerush, *Juncus roemerianus* (p. 124)

                34. Stems 8–24 inches tall, round and
                    somewhat stiff but not sharp-pointed;
                    dominant high salt marsh plant — Black Grass, *Juncus gerardii* (p. 124)

        32. Leaves numerous and not round in cross-section
            35. Leaves compound, divided into leaflets
                36. Leaves fronds of ferns
                    37. Frond leaflets irregularly lobed and
                        all leaflets joined; fruiting structures
                        borne on a separate fertile frond — Sensitive Fern, *Onoclea sensibilis* (p. 163)

                    37. Frond leaflets regularly divided
                        into secondary leaflets
                        38. Leaflets separate and borne on
                            short stalks; fruiting structures
                            (sporangia) borne on a terminal
                            spike arising from leafy fronds — Royal Fern, *Osmunda regalis* (p. 165)

48

38. Leaflets sessile, not borne on short stalks; fruiting structures (sporangia) borne on underside of leaflets — Marsh Fern, *Thelypteris thelypteroides* (p. 164)

36. Leaves not fern fronds
  39. Leaves deeply dissected into threadlike leaflets — Mock Bishop-weed, *Ptilimnium capillaceum* (p. 141)

  39. Leaflets not threadlike but simple with toothed or smooth margins
    40. Leaflet margins smooth; tidal fresh marsh plant (rare but locally common) from southern New Jersey south — Sensitive Joint Vetch, *Aeschynomene virginica* (p. 206)

    40. Leaflet margins toothed
      41. Underside of leaflets silky hairy; creeping plant of New England salt marshes — Silverweed, *Potentilla anserina* (p. 136)

      41. Underside of leaflets smooth, not hairy; plant not creeping but erect (up to 6 feet)
        42. Leaflets round-toothed, forming lobes; main lobes with sharp-pointed tips — Tall Meadow-rue, *Thalictrum pubescens* (p. 203)

        42. Leaflets not round-toothed but sharp-toothed
          43. Leaves alternately arranged; stems strongly angled or grooved — Water Parsnip, *Sium suave* (p. 218)

          43. Leaves oppositely arranged; stems not strongly angled or grooved — Beggar-ticks, *Bidens* spp. (p. 237)

35. Leaves simple, not compound
  44. Leaves grass or grasslike
    45. Stems solid and round in cross-section
      46. Leaves fleshy-textured and alternately arranged; plants of salt and brackish marshes
        47. Leaves narrowly linear; perennial with fibrous roots and creeping rhizomes — Perennial Salt Marsh Aster, *Aster tenuifolius* (p. 152)

        47. Leaves narrowly linear to narrowly oblong; annual with short taproot — Annual Salt Marsh Aster, *Aster subulatus* (p. 151)

46. Grass or grasslike leaves, not fleshy
  48. Grasslike leaves arranged in three ranks
    49. Leaves narrow, less than ⅛ inch wide, with weakly rough margins; terminal and lateral upward-branching inflorescences of spikelets in dense clusters; brackish and tidal fresh marsh plant up to 3 feet tall   Twig Rush, *Cladium mariscoides* (p. 176)

    49. Leaves wider than ⅛ inch, with very rough margins; dense clump of basal leaves present; terminal inflorescence of budlike spikelets somewhat drooping at maturity; tidal fresh marsh plant up to 6½ feet tall   Wool Grass, *Scirpus cyperinus* (p. 179)

  48. Grasslike leaves not arranged in three ranks
    50. Grasslike leaves alternately arranged and less than 6 inches long   Grass-leaved Goldenrod, *Euthamia graminifolia* (p. 155)

    50. Stiff, elongate grass leaves arranged in two ranks; leaves rough and usually rolled inwardly at distal ends; dense, terminal spike inflorescence usually four-angled at joints; upper high salt marsh plant from Massachusetts north   Stiff-leaf Quackgrass, *Agropyron pungens* (p. 101)

45. Stems not solid and round but either hollow and round or solid and triangular in cross-section
  51. Stems hollow and round
    52. Leaves distinctly three-ranked; leaf sheaths closed (not extending downward along stem); stems jointed   Three-way Sedge, *Dulichium arundinaceum* (p. 178)

    52. Leaves two-ranked; ligule present at base of leaf; leaf sheaths open, extending downward along stem, and attaching at nodes (Grasses)
      53. Leaf blades less than ⅕ inch wide, typically rolled inwardly

54. Panicle purplish, with two to six widely separated
flowering spikes and spikelets crowded and overlapping,
mostly on one side of branchlets; dominant salt and
brackish marsh plant, often forming dense mats,
giving the marsh a cowlicked appearance — Salt Hay Grass,
Spartina patens
(p. 111)

54. Panicle not purplish but much branched with numerous
spikes not widely separated and spikelets not on one
side of branchlets; plant of upper borders of salt marshes — Red Fescue, *Festuca
rubra* (p. 104)

53. Leaf blades wider than ⅕ inch; grasses not typically
forming dense mats but standing erect
55. Plant with leaves conspicuously two-ranked and
condensed terminal panicle usually less than 2 inches
long; plant growing from creeping rhizomes; commonly
intermixed with salt hay grass; may on occasion form
a dense mat in waterlogged portions of high salt marsh — Spike Grass, *Distichlis
spicata* (p. 103)

55. Plants with leaves not conspicuously two-ranked and
terminal panicle usually longer than 2 inches
56. Panicle unbranched and spikelike
57. Panicle with elongate bristles (awns) — Virginia Rye Grass,
Elymus virginicus
(p. 171)

57. Panicle (sometimes less than 2 inches long)
with short bristles (less than 1 inch long) — Foxtail Grass, *Setaria
geniculata* (p. 108)

56. Panicle branching
58. Numerous spikelets with elongate bristles,
often longer than 1 inch — Walter Millet,
Echinochloa walteri
(p. 170)

58. Spikelets without bristles or, if present, bristles
not elongate or conspicuous
59. Panicle compressed
60. Stems soft and spongy and leaves
tapering to a long, inwardly rolled tip;
common salt and brackish marsh plant — Smooth Cordgrass,
Spartina alterniflora
(p. 109)

60. Stems not soft and spongy; leaves
tapering to a tip but not rolled
inwardly

61. Panicle cylinder-shaped; creeping stems above or just below marsh surface; stems usually 16 inches tall or less at maturity; brackish and tidal fresh marsh plant — Creeping Bent Grass, *Agrostis stolonifera* var. *compacta* (p. 102)

61. Panicle narrow, branches spreading during flowering; stems usually over 24 inches tall at maturity; tidal fresh marsh plant — Reed Canary Grass, *Phalaris arundinacea* (p. 173)

59. Panicle open and spreading
    62. Spikelets crowded, mostly on one side of branchlets (rachis); leaf edges rough — Slough Grass, *Spartina pectinata* (p. 112)

    62. Spikelets not crowded on one side of rachis
        63. Panicle erect
            64. Few spikelets located mostly at tips of branchlets; plant usually forms dense clumps at upper edges of salt marshes — Switchgrass, *Panicum virgatum* (p. 105)

            64. Many spikelets located along length of branchlets (one species with spikelets only on outer half of branchlets); leaf blades smooth and often inwardly rolled; salt marsh plant — Alkali Grasses, *Puccinellia* spp. (p. 107)

        63. Panicle not entirely erect but somewhat drooping; tidal fresh marsh plant — Bluejoint, *Calamagrostis canadensis* (p. 169)

51. Stems solid and triangular in cross-section; leaves three-ranked; leaf sheaths closed (not extending downward along stem); no ligule present (Sedges)
    65. Stems weakly triangular-roundish, more triangular toward base
        66. Leaves narrow, less than ⅛ inch wide, with weakly rough margins; terminal and lateral upward-branching inflorescences of spikelets in dense clusters; brackish and tidal fresh marsh plant up to 3 feet tall — Twig Rush, *Cladium mariscoides* (p. 176)

66. Leaves wider than ⅛ inch with very rough margins; dense clump of basal leaves present; terminal inflorescence of budlike spikelets somewhat drooping at maturity; tidal fresh marsh plant up to 6½ feet tall — Wool Grass, *Scirpus cyperinus* (p. 179)

65. Stems strongly triangular
    67. Stems thin
        68. Forming dense clumps or tussocks; two kinds of flowers borne on a single plant, male (staminate) flowered spikelets above and female (pistillate) flowered spikelets below; tidal fresh marsh plant — Tussock Sedge, *Carex stricta* (p. 175)

        68. Not tussock-forming
            69. Two kinds of spikelets (staminate and pistillate) present on same plant; deeply drooping spikelets on slender peduncles; salt marsh plant from Massachusetts north — Salt Marsh Sedge, *Carex paleacea* (p. 114)

            69. One kind of spikelet present
                70. Terminal erect spike of three to nine somewhat egg-shaped clusters, each subtended by a bristlelike bract; brackish and tidal fresh marsh plant — Marsh Straw Sedge, *Carex hormathodes* (p. 113)

                70. Terminal inflorescence of numerous budlike spikelets borne on stalks, some erect and others slightly drooping; inflorescence subtended by two or three leaflike bracts; upper salt marsh plant from Long Island south — Salt Marsh Fimbristylis, *Fimbristylis castanea* (p. 118)

65. Stems thick and stout
    71. Spikelets budlike and mostly drooping, some sessile; tidal fresh marsh plant — River Bulrush, *Scirpus fluviatilis* (p. 180)

    71. Spikelets budlike or flattened and not mostly drooping but mostly erect or sessile
        72. Spikelets resembling brown, scaly buds, subtended by mostly erect leafy bracts — Salt Marsh Bulrush, *Scirpus robustus* (p. 121)

72. Spikelets flattened, not budlike, and subtended by
drooping or weakly erect leafy bracts
  73. Tidal fresh marsh plant   Umbrella Sedge,
Cyperus strigosus
(p. 177)

  73. Salt and brackish marsh plant   Nuttall's Cyperus,
Cyperus filicinus
(p. 115)

44. Leaves not grasslike
  74. Stem square in cross-section; leaves oppositely arranged
    75. Stems and leaves minty-smelling when crushed   Wild Mint, *Mentha
arvensis* (p. 226)

    75. Stems and leaves not minty-smelling
      76. Branches distinctly arching, often rooting at
tips; stems four-to-six-angled; leaves with
smooth margins, sometimes in whorls of
threes or fours   Water Willow,
Decodon verticillatus
(p. 215)

      76. Branches not distinctly arching or rooted at tips
        77. Leaves stalked (petioled)
          78. Stems very hairy and rarely branched;
leaves mostly rough or hairy above and
hairy below; plant of upper edges of
salt marshes and also in brackish and
tidal fresh marshes   American Germander,
Teucrium canadense
(p. 147)

          78. Stems smooth or fine hairy on angles
and much branched to unbranched;
leaves mostly smooth above and below;
tidal fresh marsh plant   Mad-dog Skullcap,
Scutellaria lateriflora
(p. 227)

        77. Leaves not stalked (sessile)
          79. Leaf margins smooth; leaf bases heart-
shaped and somewhat clasping stem;
leaves sometimes in whorls of threes   Purple Loosestrife,
Lythrum salicaria
(p. 216)

          79. Leaf margins toothed
            80. Stems usually fine hairy; leaf teeth
beginning just below middle of
leaf; lower leaf margin smooth   Water Horehound,
Lycopus virginicus
(p. 225)

80. Stems smooth; leaf teeth along
    entire margin .... Square-stemmed
                                                    Monkeyflower,
                                                    Mimulus ringens
                                                    (p. 229)
74. Stems not square in cross-section
  81. Sap milky
    82. Leaves oppositely arranged and smooth-margined ... Swamp Milkweed,
                                                    Asclepias incarnata
                                                    (p. 222)

    82. Leaves alternately arranged and irregularly toothed ... Cardinal Flower,
                                                    Lobelia cardinalis
                                                    (p. 234)

  81. Sap not milky
    83. Stem with jointed nodes (where leaves join stem)
      84. Stem smooth, with sheaths above each joint
        85. Tidal fresh marsh plant .... Water Smartweed,
                                                    Polygonum punctatum
                                                    (p. 199)

        85. Salt marsh plant, along upper borders ... Bushy Knotweed,
                                                    *Polygonum
                                                    ramosissimum*
                                                    (p. 126)

      84. Stem grooved, without sheaths .... Swamp Dock, *Rumex
                                                    verticillatus* (p. 201)

    83. Stem lacking jointed nodes
      86. Leaves arranged in whorls of threes or fours ... Purple Joe-Pye-weed,
                                                    *Eupatoriadelphus
                                                    purpureus* (p. 238)

      86. Leaves not arranged in whorls
        87. Leaves oppositely arranged
          88. Leaves clasping stem and joined at
              bases .... Boneset, *Eupatorium
                                                    perfoliatum* (p. 239)

          88. Leaves not joined at bases
            89. Leaves with three to five deep lobes
                and rough on both sides .... Giant Ragweed,
                                                    Ambrosia trifida
                                                    (p. 235)

            89. Leaves not three-to-five-lobed or
                rough on both sides
              90. Leaves very narrowly linear
                  (almost grasslike) and not
                  toothed .... Purple Gerardia,
                                                    Agalinis purpurea
                                                    (p. 148)

90. Leaves not linear or only upper leaves linear
    91. Leaves egg-shaped to oblong, smooth-margined, and sessile
        92. Leaves with round tips and somewhat heart-shaped bases; tidal fresh marsh plant

Marsh St. John's-wort, *Triadenum virginicum* (p. 213)

        92. Leaves somewhat egg-shaped, with tapered tips and narrowing near stem; salt and brackish marsh plant
            93. Leaves widest below middle; upper leaves linear; annual

Annual Marsh Pink, *Sabatia stellaris* (p. 146)

            93. Leaves narrowly egg-shaped; perennial; from Long Island south

Perennial Marsh Pink, *Sabatia dodecandra* (p. 145)

    91. Leaves not egg-shaped to oblong but lance-shaped
        94. Leaves stalked (petioled), somewhat broadly lance-shaped, tapering to a long point distally; tidal fresh marsh plant

False Nettle, *Boehmeria cylindrica* (p. 197)

        94. Leaves not stalked or only young leaves stalked
            95. Salt and brackish marsh plant
                96. Leaves widest below middle, upper leaves linear; annual

Annual Marsh Pink, *Sabatia stellaris* (p. 146)

                96. Upper leaves not linear; perennial; from Long Island south

Perennial Marsh Pink, *Sabatia dodecandra* (p. 145)

            95. Tidal fresh marsh plant
                97. Plant less than 12 inches tall at maturity

Overlooked Hedge Hyssop, *Gratiola neglecta* (p. 228)

                97. Plant much taller than 12 inches at maturity
                    98. Leaves with heart-shaped bases, somewhat clasping stem (sometimes in whorls of threes); also occurring along edges of salt and brackish marshes

Purple Loosestrife, *Lythrum salicaria* (p. 216)

                    98. Leaves toothed and not clasping stem

Bur Marigold, *Bidens laevis* (p. 237)

87. Leaves alternately arranged
    99. Leaves aromatic when crushed   Annual Salt Marsh Fleabane, *Pluchea purpurascens* (p. 157)

    99. Leaves not aromatic
        100. Leaves usually three-lobed and toothed
            101. Leaves smooth above and fine hairy below; upper leaves may be egg-shaped without lobes   Rose Mallow, *Hibiscus moscheutos* (p. 138)

            101. Leaves hairy on both sides
                102. Stems and leaves velvety smooth hairy   Marsh Mallow, *Althaea officinalis* (p. 137)

                102. Stems and leaves hairy and rough, not velvety smooth   Seashore Mallow, *Kosteletzkya virginica* (p. 139)

        100. Leaves not three-lobed
            103. Leaves somewhat fleshy
                104. Leaves linear, grasslike, and weakly toothed   Perennial Salt Marsh Aster, *Aster tenuifolius* (p. 152)

                104. Leaves not linear
                    105. Leaves soft, coarse-toothed, and on long petioles; tidal fresh marsh plant   Jewelweed, *Impatiens capensis* (p. 212)

                    105. Leaves firm and sessile, not on petioles; salt and brackish marsh plant
                        106. Leaves oblong and without callous-tipped teeth   Seaside Goldenrod, *Solidago sempervirens* (p. 158)

                        106. Leaves sharply toothed with callous-tipped teeth   Fireweed, *Erechtites hieracifolia* (p. 154)
            103. Leaves not fleshy
                107. Leaves sharply toothed with callous-tipped teeth (sometimes fleshy)   Fireweed, *Erechtites hieracifolia* (p. 154)

107. Leaves not sharply toothed with callous-tipped teeth
    108. Leaves, after connecting to stem, extending downward as wings along stem; tidal fresh marsh plant — Sneezeweed, *Helenium autumnale* (p. 240)

    108. Leaves not forming wings along stem
        109. Leaves linear, grasslike
            110. Leaves somewhat fleshy — Perennial Salt Marsh Aster, *Aster tenuifolius* (p. 152)

            110. Leaves not somewhat fleshy — Grass-leaved Goldenrod, *Euthamia graminifolia* (p. 155)

        109. Leaves not grasslike
            111. Leaves slightly clasping stem — New York Aster, *Aster novi-belgii* (p. 236)

            111. Leaves not clasping stem
                112. Leaves on long stalks (petioles)
                    113. Leaves lance-shaped; brackish and tidal fresh marsh plant — Water Hemp, *Amaranthus cannabinus* (p. 132)

                    113. Leaves somewhat heart-shaped tapering to a point distally, stalks sheathing stem at base; tidal fresh marsh plant — Lizard's Tail, *Saururus cernuus* (p. 193)

                112. Leaves not on long stalks (petioles)
                    114. Leaves spoon-shaped or somewhat oval, both basal and alternately arranged; brackish and tidal fresh marsh plant less than 20 inches tall — Water Pimpernel, *Samolus parviflorus* (p. 143)

                    114. Leaves not spoon-shaped or somewhat oval but lance-shaped; leaves only alternately arranged; tidal fresh marsh plant
                        115. Leaf margins finely toothed; leaves rough

| hairy above and thin hairy below | New York Ironweed, *Vernonia noveboracensis* (p. 242) |

115. Leaf margins smooth; leaves smooth above and below — Seedbox, *Ludwigia alternifolia* (p. 217)

Key F   Tall Herbaceous Plants

1. Grass or grasslike plants
    2. Stems strongly or weakly triangular in cross-section; leaves three-ranked
        3. Stems weakly triangular or somewhat rounded in cross-section, more triangular near base of plant; dense clump of basal leaves present — Wool Grass, *Scirpus cyperinus* (p. 179)

        3. Stems strongly triangular in cross-section
            4. Stems without apparent leaves and deeply concave in cross-section; brackish marsh plant — Olney Three-square, *Scirpus americanus* (p. 119)

            4. Stems with leaves; tidal fresh marsh plant — River Bulrush, *Scirpus fluviatilis* (p. 180)

    2. Stems not triangular
        5. Stems leafless and round in cross-section — Soft-stemmed Bulrush, *Scirpus validus* (p. 122)

        5. Stems not entirely leafless
            6. Elongate (usually more than 4 feet), erect, narrow, and flattened basal leaves only, forming long sheaths along stem before branching outward in an alternately arranged fashion — Cattails, *Typha* spp. (p. 99)

            6. Leaves not so
                7. Leaves three-ranked; dense clump of basal leaves present; terminal inflorescence of stalked, budlike spikelets, somewhat drooping at maturity — Wool Grass, *Scirpus cyperinus* (p. 179)

7. Leaves two-ranked (Grasses)
   8. Panicle with two types of spikelets (male spikelets below female spikelets); common tidal fresh marsh plant   Wild Rice, *Zizania aquatica* (p. 174)

   8. Panicle with one type of spikelets
     9. Panicle subtended by tuft of silky, light brown hairs; feathery appearance when mature   Common Reed, *Phragmites australis* (p. 106)

     9. Panicle not subtended by tuft of silky hairs
       10. Spikelets with elongate bristles, often larger than 1 inch   Walter Millet, *Echinochloa walteri* (p. 170)

       10. Spikelets lacking elongate bristles
         11. Spikelets crowded and overlapping, mostly on one side of branchlets (rachis)
           12. Leaf margins rough; inflorescence branches spreading or ascending
             13. Principal leaf blades ⅓–⅖ inch wide and leaf margins strongly toothed   Big Cordgrass, *Spartina cynosuroides* (p. 110)

             13. Principal leaf blades ⅕–⅖ inch wide, very long, and rolled inward when dry   Slough Grass, *Spartina pectinata* (p. 112)

           12. Leaf margins smooth or only slightly rough; inflorescence branches appressed; plant dominating creek banks in salt and brackish marshes   Smooth Cordgrass, *Spartina alterniflora* (p. 109)

         11. Spikelets not crowded on one side of branchlets (rachis); few spikelets located mostly at tips of branchlets; plant usually forming dense clumps at upper edges of salt marshes   Switchgrass, *Panicum virgatum* (p. 105)

1. Plants not grass or grasslike
   14. Sap milky
       15. Leaves oppositely arranged and smooth-margined — Swamp Milkweed, *Asclepias incarnata* (p. 222)

       15. Leaves alternately arranged and irregularly toothed — Cardinal Flower, *Lobelia cardinalis* (p. 234)
   14. Sap not milky
       16. Leaves compound, divided into many leaflets, and alternately arranged
           17. Leaflets toothed
               18. Leaflets strongly toothed and linear or lance-shaped; stem strongly angled or grooved — Water Parsnip, *Sium suave* (p. 218)

               18. Leaflets round-toothed, forming lobes, main lobes having sharp-pointed tips — Tall Meadow-rue, *Thalictrum pubescens* (p. 203)

           17. Leaflets not toothed; tidal fresh marsh plant from southern New Jersey south — Sensitive Joint Vetch, *Aeschynomene virginica* (p. 206)

       16. Leaves simple, not divided into leaflets
           19. Leaves arranged in whorls of threes or fours — Purple Joe-Pye-weed, *Eupatoriadelphus purpureus* (p. 238)

           19. Leaves not arranged in whorls
               20. Leaves oppositely arranged
                   21. Leaves clasping stem and joined at bases — Boneset, *Eupatorium perfoliatum* (p. 239)

                   21. Leaves not joined at bases; leaves three-to-five-lobed — Giant Ragweed, *Ambrosia trifida* (p. 235)

               20. Leaves alternately arranged
                   22. Leaves usually three-lobed and toothed
                       23. Leaves smooth above and fine hairy below; upper leaves may be egg-shaped without lobes — Rose Mallow, *Hibiscus moscheutos* (p. 138)

23. Leaves hairy on both sides	Seashore Mallow, *Kosteletzkya virginica* (p. 139)
22. Leaves not three-lobed	
24. Leaves sharply toothed with callous-tipped teeth (sometimes fleshy)	Fireweed, *Erechtites hieracifolia* (p. 154)
24. Leaves not sharply toothed with callous-tipped teeth	
25. Leaves, after connecting to stem, extending downward along stem as wings	Sneezeweed, *Helenium autumnale* (p. 240)
25. Leaves not forming wings along stem	
26. Leaves entire; brackish and tidal fresh marsh plant	Water Hemp, *Amaranthus cannabinus* (p. 132)
26. Leaves toothed, not entire; tidal fresh marsh plant	
27. Leaves egg-shaped, soft, almost fleshy	Jewelweed, *Impatiens capensis* (p. 212)
27. Leaves not egg-shaped and not soft but lance-shaped (tapering at both ends) and firm	
28. Leaves irregularly toothed	Cardinal Flower, *Lobelia cardinalis* (p. 234)
28. Leaves regularly fine-toothed and somewhat rough above	New York Ironweed, *Vernonia noveboracensis* (p. 242)

Key G Marsh Plants in Flower

1. Flowering vine or woody shrub	
2. Flowering vine	
3. Flowers pealike, fragrant, and brown or purple	Ground-nut, *Apios americana* (p. 208)
3. Flowers not pealike or fragrant	
4. Flowers six-petaled, white or greenish yellow	Wild Yam, *Dioscorea villosa* (p. 191)

4. Flowers not six-petaled
  5. White or pink flowers in heads borne in petioled clusters — Climbing Hempweed, *Mikania scandens* (p. 241)

  5. Flowers tubular, not in heads
    6. Small white or yellowish bell-shaped flowers (1/10–1/5 inch long) in sessile clusters — Dodder, *Cuscuta gronovii* (p. 224)

    6. Larger white or pink tubular (funnel-shaped) flowers (1½–3 inches long) borne usually singly on long peduncles — Hedge Bindweed, *Calystegia sepium* (p. 223)

2. Flowering woody shrub
  7. Branches bearing thorns; pink, five-petaled flowers — Swamp Rose, *Rosa palustris* (p. 205)

  7. Branches without thorns
    8. Leaves aromatic (bayberry scent)
      9. Plant occurring from Long Island north along edges of salt marshes — Sweet Gale, *Myrica gale* (p. 125)

      9. Plant occurring from southern New Jersey south along edges of salt and brackish marshes and in tidal fresh marshes — Wax Myrtle, *Myrica cerifera* (p. 195)

    8. Leaves not aromatic
      10. Small red flowers in short clusters; leaves not present when flowering — Red Maple, *Acer rubrum* (p. 211)

      10. Flowers not red
        11. Flowers purplish, numerous, borne on dense, spikelike inflorescence — False Indigo, *Amorpha fruticosa* (p. 207)

        11. Flowers not purplish
          12. Flowers pink, five-petaled, and bell-shaped; stems arching branches — Water Willow, *Decodon verticillatus* (p. 215)

          12. Flowers not pink
            13. Greenish or greenish white flowers
              14. Flowers borne in branching clusters from leaf axils — Poison Ivy, *Toxicodendron radicans* (p. 209)

14. Flowers borne in leafy terminal inflorescence; common salt marsh shrub — High-tide Bush, *Iva frutescens* (p. 156)

13. White flowers
    15. Flowers borne in dense, ball-like clusters — Buttonbush, *Cephalanthus occidentalis* (p. 230)

    15. Flowers not borne in ball-like clusters
        16. Flowers four-petaled, numerous, and borne in branching terminal inflorescences at tips of branches — Silky Dogwood, *Cornus amomum* (p. 219)

        16. Flowers not four-petaled or, if four-petaled, not borne in branching terminal inflorescences
            17. Flowers fragrant, small, five-petaled, numerous, borne on dense spikelike inflorescences; tidal freshwater wetland plant — Sweet Pepperbush, *Clethra alnifolia* (p. 220)

            17. Flowers not fragrant
                18. Flowers with four, five, or six petals or lobes
                    19. Borne singly or in loose clusters along the stem; four to six petals — Common Winterberry, *Ilex verticillata* (p. 210)

                    19. Borne in dense clusters; usually five-lobed
                        20. Leaves compound, divided into five to eleven, usually seven, leaflets — Common Elder, *Sambucus canadensis* (p. 232)

                        20. Leaves simple, not compound — Northern Arrowwood, *Viburnum recognitum* (p. 233)

                18. Flowers in heads, lacking petals
                    21. Leaves fleshy and oppositely arranged — High-tide Bush, *Iva frutescens* (p. 156)

                    21. Leaves alternately arranged and not fleshy — Groundsel Tree, *Baccharis halimifolia* (p. 153)

1. Flowering, erect herbaceous plant
22. Fleshy leaves and/or stems
   23. Flowers yellow or orange
      24. Orange tubular flowers with brown spots borne singly on long petioles — Jewelweed, *Impatiens capensis* (p. 212)

    24. Yellow flowers
      25. Irislike flowers; leaves sword-shaped — Yellow Flag, *Iris pseudacorus* (p. 192)

      25. Flowers not irislike; leaves not sword-shaped
        26. Single, large flower (1½–2 inches wide) with five or six petals, borne on a stalk — Spatterdock, *Nuphar luteum* (p. 202)

        26. Flowers numerous, not solitary
          27. Small yellow flowers borne on fleshy stalk (spadix); basal leaves fleshy and egg-shaped — Golden Club, *Orontium aquaticum* (p. 182)

          27. Numerous small flowers with seven to seventeen rays borne in heads on terminal inflorescences — Seaside Goldenrod, *Solidago sempervirens* (p. 158)

   23. Flowers not yellow or orange
     28. Purple, light blue, or pink flowers
      29. Numerous small purplish tubular flowers borne on a terminal spike — Pickerelweed, *Pontederia cordata* (p. 188)

      29. Flowers not borne on a terminal spike
        30. Irislike flowers; leaves sword-shaped — Blue Flag, *Iris versicolor* (p. 192)

        30. Flowers not irislike; leaves not sword-shaped
          31. Flowers daisylike with fifteen to twenty-five petallike rays borne in heads — Perennial Salt Marsh Aster, *Aster tenuifolius* (p. 152)

          31. Flowers not daisylike, not in heads
            32. Flowers four-petaled — Sea Rocket, *Cakile edentula* (p. 135)

32. Flowers not four-petaled
   33. Flowers six-petaled, star-shaped, and light blue; basal leaves heart-shaped or kidney-shaped — Mud Plantain, *Heteranthera reniformis* (p. 187)

   33. Pinkish flowers with five petals or lobes; leaves not basal or heart-shaped or kidney-shaped
     34. Tubular flowers with five lobes
       35. Small pinkish to whitish flowers (up to ¼ inch long) with short tubes borne singly in leaf axils — Sea Milkwort, *Glaux maritima* (p. 142)

       35. Larger flowers (½ inch wide) with long tubes borne in two to five pairs on stalks — Seaside Gerardia, *Agalinis maritima* (p. 148)

     34. Flowers not tubular but with five separate petals; flowers small (⅙ inch wide) — Salt Marsh Sand Spurrey, *Spergularia marina* (p. 133)

28. Flowers not purple, light blue, or pink
   36. White flowers
     37. Flowers in heads, with or without petallike rays
       38. Flowers daisylike with fifteen to twenty-five petallike rays — Perennial Salt Marsh Aster, *Aster tenuifolius* (p. 152)

       38. Flowers without rays and not daisylike; sharply toothed leaves with callous tips — Fireweed, *Erechtites hieracifolia* (p. 154)

     37. Flowers not in heads
       39. Small four-petaled flowers
         40. Tidal fresh marsh plant up to 4 inches tall; flowers sometimes greenish white — Pygmyweed, *Crassula aquatica* (p. 204)

         40. Salt marsh plant 8–12 inches tall; fleshy leaves with mild horseradish taste — Sea Rocket, *Cakile edentula* (p. 135)

       39. Flowers not four-petaled
         41. Flowers six-petaled and star-shaped; tidal fresh marsh plant — Mud Plantain, *Heteranthera reniformis* (p. 187)

         41. Flowers not six-petaled; salt marsh plants

42. Flowers five-petaled; triangular structures (stipules) at leaf bases — Salt Marsh Sand Spurrey, *Spergularia marina* (p. 133)

42. Flowers five-lobed, forming short tube at base — Sea Milkwort, *Glaux maritima* (p. 142)

36. Flowers not white
43. Red flowers five-lobed, forming short tube at base — Sea Milkwort, *Glaux maritima* (p. 142)

43. Flowers not red
44. Flowers borne on fleshy stalk (spadix) and covered by a hoodlike structure (spathe); spathe spotted and striped purplish and greenish; plant foul-smelling (skunklike odor) — Skunk Cabbage, *Symplocarpus foetidus* (p. 184)

44. Flowers not covered by hoodlike spathe and plant not foul-smelling
45. Flowers not apparent, enclosed within a fleshy leaflike structure (spathe) — Arrow Arum, *Peltandra virginica* (p. 183)

45. Flowers visible, not hidden
46. Flowers inconspicuous and borne on leafless stem or stem with sheathing basal leaves
47. Entire plant apparently leafless
48. Solitary stem with woody core — Perennial Glasswort, *Salicornia virginica* (p. 129)

48. Branching stem without woody core
49. Stem greater than ⅕ inch wide — Bigelow's Glasswort, *Salicornia bigelovii* (p. 128)

49. Stem less than ⅕ inch wide — Common Glasswort, *Salicornia europaea* (p. 128)

47. Basal leaves present
50. Plants tall, greater than 4 feet in height; basal leaves forming long sheaths along flowering stem before branching outward in an alternately arranged fashion — Cattails, *Typha* spp. (p. 99)

50. Plants not tall, much less than 4 feet in height
   51. Leaves with sheaths

Seaside Arrow Grass, *Triglochin maritimum* (p. 100)

   51. Leaves without sheaths

Seaside Plantain, *Plantago maritima* (p. 150)

46. Flowers small and borne on leafy stems
   52. Leaves arrowhead-shaped

Marsh Orach, *Atriplex patula* (p. 127)

   52. Leaves linear, not arrowhead-shaped

Sea Blite, *Suaeda linearis* (p. 131)

22. Leaves and/or stems not fleshy
   53. Orange flowers
      54. Flowers tubular with reddish brown spots, borne singly on long stalks

Jewelweed, *Impatiens capensis* (p. 212)

      54. Flowers not borne singly in leaf axils but borne in terminal inflorescence; milky sap; plant of southern brackish and tidal fresh marshes, from southern New Jersey south

Red Milkweed, *Asclepias lanceolata* (p. 222)

   53. Flowers not orange
      55. Pink flowers
         56. Stems spiny
            57. Flowers with five lobes; leaves narrowly arrowhead-shaped

Arrow-leaved Tearthumb, *Polygonum sagittatum* (p. 200)

            57. Flowers with four lobes; leaves broadly arrowhead-shaped

Halberd-leaved Tearthumb, *Polygonum arifolium* (p. 198)

         56. Stems not spiny
            58. Stems with four to six angles
               59. Bell-shaped flowers with five petals borne in clusters in axils of upper leaves; stems with four to six angles and distinctly arching

Water Willow, *Decodon verticillatus* (p. 215)

59. Flowers irregular with broad lower lip, upper lip absent, borne in dense terminal spike; stems four-angled (square in cross-section), hairy, and erect

American Germander, *Teucrium canadense* (p. 147)

58. Stems not four-to-six-angled
    60. Stems less than 2 inches long and creeping; inconspicuous three-petaled flowers; mat-forming plant of mud flats along tidal fresh waters

American Waterwort, *Elatine americana* (p. 214)

    60. Stems much greater than 4 inches tall and erect, not mat-forming
        61. Stems with milky sap; flowers borne in terminal inflorescence (umbel)

Swamp Milkweed, *Asclepias incarnata* (p. 222)

        61. Stems without milky sap
            62. Flowers in heads
                63. Leaves strongly aromatic and alternately arranged; salt and brackish marsh plant

Salt Marsh Fleabane, *Pluchea purpurascens* (p. 157)

                63. Leaves not aromatic or only weakly so and arranged in whorls; tidal fresh marsh plant

Joe-Pye-weeds, *Eupatoriadelphus* spp. (p. 238)

            62. Flowers not in heads but with five or more petals
                64. Eight to twelve petals

Perennial Marsh Pink, *Sabatia dodecandra* (p. 145)

                64. Five petals
                    65. Petals joined at bases, forming tubular flowers

Purple Gerardia, *Agalinis purpurea* (p. 148)

                    65. Petals separate, not forming tubular flowers
                        66. Flowers larger than 4 inches wide

Rose Mallow, *Hibiscus moscheutos* (p. 138)

                        66. Flowers less than 4 inches wide
                            67. Stems smooth

69

68. Flowers with yellow center and ¾–1½ inches wide; salt and brackish marsh plant

Annual Marsh Pink, *Sabatia stellaris* (p. 146)

68. Flowers without yellow center and up to ⅕ inch wide; tidal fresh marsh plant

Marsh St. John's-wort, *Triadenum virginicum* (p. 213)

67. Stems hairy
  69. Stems rough; flowers 1½–2½ inches wide

Seashore Mallow, *Kosteletzkya virginica* (p. 139)

  69. Stems velvety soft; flowers 1–1½ inches wide

Marsh Mallow, *Althaea officinalis* (p. 137)

55. Flowers not pink
  70. Yellow flowers
    71. Irislike flowers with sword-shaped leaves

Yellow Flag, *Iris pseudacorus* (p. 192)

    71. Flowers not irislike and leaves not sword-shaped
      72. Leaves compound, divided into separate leaflets
        73. Flowers pealike with two lips, short tube, and red veins

Sensitive Joint Vetch, *Aeschynomene virginica* (p. 206)

        73. Flowers not pealike with red veins
          74. Flowers in dense heads, lacking petals; tidal fresh marsh plants

Beggar-ticks, *Bidens* spp. (p. 237)

          74. Single five-petaled flower borne on stalk from creeping runner; salt and brackish marsh plant

Silverweed, *Potentilla anserina* (p. 136)

      72. Leaves simple, not compoundly divided
        75. Leaves aromatic; flowers borne on elongate, fleshy, fingerlike appendage (spadix)

Sweet Flag, *Acorus calamus* (p. 181)

        75. Leaves not aromatic and flowers not borne on spadix
          76. Plants usually less than 12 inches tall
            77. Flowers five-petaled with dense, conelike head; leaves mostly basal, oval heart-shaped to kidney-shaped;

brackish and tidal fresh marsh plant common on mud flats	Seaside Crowfoot, *Ranunculus cymbalaria* (p. 134)
77. Tubular flowers with five lobes; tidal fresh marsh plant	
78. Tube yellow and lobes white	Overlooked Hedge Hyssop, *Gratiola neglecta* (p. 228)
78. Tube and lobes bright yellow; plant up to 16 inches tall	Golden-pert, *Gratiola aurea* (p. 228)
76. Plants much taller than 12 inches	
79. Flowers in heads, with or without petallike rays	
80. Flowers not daisylike, borne on dense terminal inflorescences	
81. Leaves somewhat fleshy	Seaside Goldenrod, *Solidago sempervirens* (p. 158)
81. Leaves grasslike, not fleshy	Grass-leaved Goldenrod, *Euthamia graminifolia* (p. 155)
80. Flowers daisylike, not borne on dense terminal inflorescences	
82. Flowers with ten to twenty wedge-shaped, drooping rays with three-lobed tips	Sneezeweed, *Helenium autumnale* (p. 240)
82. Flowers with seven to eight rays, without lobed tips	Bur Marigold, *Bidens laevis* (p. 237)
79. Flowers not in heads	
83. Four-petaled flowers borne singly in leaf axils	Seedbox, *Ludwigia alternifolia* (p. 217)
83. Flowers inconspicuous, borne on slender spikes	Water Hemp, *Amaranthus cannabinus* (p. 132)
70. Flowers not yellow	
84. Purple or bluish flowers	
85. Irislike flowers with somewhat fleshy sword-shaped leaves	Blue Flag, *Iris versicolor* (p. 192)
85. Flowers not irislike	
86. Stems square or many-angled	

87. Stems distinctly arching and with four to six angles;
    pinkish purple bell-shaped flowers with five petals,
    borne in clusters in axils of upper leaves

    Water Willow,
    Decodon verticillatus
    (p. 215)

87. Stems not arching but erect
    88. Stems and leaves aromatic, with strong minty odor;
        numerous small tubular flowers (light blue,
        lavender, or white) in dense ball-like clusters in
        leaf axils

        Wild Mint, *Mentha
        arvensis* (p. 226)

    88. Stems and leaves not aromatic and flowers not in
        ball-like clusters
        89. Flowers with five or six petals, borne in dense,
            leafy terminal spikes

            Purple Loosestrife,
            Lythrum salicaria
            (p. 216)

        89. Tubular flowers with two lips, one upper and
            one lower; stems square
            90. Solitary flowers borne on stalks from leaf
                axils
                91. Flower stalks long, ⅗–2⅖ inches;
                    upper lip of flower two-lobed, lower lip
                    three-lobed

                    Square-stemmed
                    Monkeyflower,
                    Mimulus ringens
                    (p. 229)

                91. Flower stalks short, less than ⅗ inch
                    long; upper lip of flower curved,
                    hoodlike in appearance

                    Common Skullcap,
                    Scutellaria galericulata
                    var. *epilobiifolia*
                    (p. 227)

            90. Flowers not solitary but numerous, borne
                on one side of axillary and terminal
                inflorescences, each usually subtended by
                small lance-shaped leaves, with upper lip
                of flower curved, hoodlike in appearance

                Mad-dog Skullcap,
                Scutellaria lateriflora
                (p. 227)

86. Stems not square or many-angled
    92. Flowers borne on single, many-branched inflorescence
        arising from basal leaves

        Sea Lavender,
        Limonium nashii
        (p. 144)

92. Flowers not borne on separate inflorescence arising from basal leaves
  93. Flowers with five petals or four to seven lobes
    94. Five-petaled flowers borne in terminal and axillary inflorescences; tidal fresh marsh plant — Marsh St. John's-wort, *Triadenum virginicum* (p. 213)

    94. Four-to-seven-lobed tubular flowers (petals joined at base to form tube) borne singly in leaf axils; salt and brackish marsh plants
      95. Five-lobed flowers longer than ½ inch, with broad, funnel-shaped tube — Purple Gerardia, *Agalinis purpurea* (p. 148)

      95. Four-to-seven-lobed flowers less than ½ inch long, with slender tube, flowers also white; occurring from southern New Jersey south — Salt Marsh Loosestrife, *Lythrum lineare* (p. 216)

  93. Flowers in heads, not four-to-seven-lobed and tubular or five-petaled
    96. Flowers with petallike rays
      97. Rays prominent; flowers daisylike
        98. Leaves linear and fleshy; salt and brackish marsh plant — Perennial Salt Marsh Aster, *Aster tenuifolius* (p. 152)

        98. Leaves not linear or fleshy; tidal fresh marsh plant — New York Aster, *Aster novi-belgii* (p. 236)

      97. Rays very short, almost inconspicuous; salt and brackish marsh plant — Annual Salt Marsh Aster, *Aster subulatus* (p. 151)

    96. Flowers without rays, deep purple; tidal fresh marsh plant — New York Ironweed, *Vernonia noveboracensis* (p. 242)

84. Flowers not purple or blue
  99. White flowers
    100. Plants less than 4 inches tall

101. Flowers borne at end of naked stalk in a single
cluster, resembling a hatpin
                                                        Parker's Pipewort,
                                                        Eriocaulon parkeri
                                                        (p. 186)

101. Flowers and stalk not resembling a hatpin
    102. Five-lobed tubular flowers borne singly on
        stalks
                                                        Mudwort, *Limosella*
                                                        subulata (p. 149)
        102. Flowers borne in terminal inflorescence (umbel)   Eastern Lilaeopsis,
                                                        Lilaeopsis chinensis
                                                        (p. 140)

100. Plants taller than 4 inches in height
    103. Stems jointed; flowers small
        104. Numerous flowers borne on erect, dense,
            terminal spikelike inflorescence, usually
            nodding at tip; flowers fragrant and without
            petals or petallike lobes
                                                        Lizard's Tail, *Saururus*
                                                        cernuus (p. 193)

        104. Flowers not borne on dense, erect terminal
            spike; flowers with petallike lobes and not
            fragrant
            105. Stems spiny
                                                        Arrow-leaved
                                                        Tearthumb,
                                                        Polygonum sagittatum
                                                        (p. 200)

            105. Stems not spiny
                                                        Water Smartweed,
                                                        Polygonum punctatum
                                                        (p. 199)
    103. Stems not jointed
        106. Three-petaled flowers usually borne in whorls
            of two to fifteen on a single elongate stalk.
            107. Leaves egg-shaped
                                                        Northern Water
                                                        Plantain, *Alisma*
                                                        plantago-aquatica
                                                        (p. 167)

            107. Leaves not egg-shaped
                108. Leaves arrowhead-shaped
                                                        Big-leaved Arrowhead,
                                                        Sagittaria latifolia
                                                        (p. 168)

                108. Leaves linear and flattened
                                                        Owl-leaf Arrowhead,
                                                        Sagittaria subulata
                                                        (p. 168)
        106. Flowers not three-petaled
            109. Compound leaves divided into leaflets;
                flowers arranged in terminal inflorescences

74

110. Leaves divided into threadlike leaflets; inflorescence umbel — Mock Bishop-weed, *Ptilimnium capillaceum* (p. 141)

110. Leaves not divided into threadlike leaflets
    111. Leaves divided into round-toothed lobes with sharp-pointed tips; inflorescence panicle — Tall Meadow-rue, *Thalictrum pubescens* (p. 203)

    111. Leaves divided into seven to seventeen lance-shaped leaflets; inflorescence umbel — Water Parsnip, *Sium suave* (p. 218)

109. Leaves simple, not compound
    112. Flowers in heads, with or without petallike rays
        113. Flowers daisylike with petallike rays
            114. Salt and brackish marsh plant; fifteen to twenty-five petallike rays; leaves linear and not toothed — Perennial Salt Marsh Aster, *Aster tenuifolius* (p. 152)

            114. Tidal fresh marsh plant; twenty to forty petallike rays; leaves lance-shaped and toothed, sometimes not toothed — Lowland White Aster, *Aster simplex* (p. 236)

        113. Petallike rays absent; flowers not daisylike
            115. Heads ball-shaped, roundish; linear leaves often somewhat triangular in cross-section — Great Bur-reed, *Sparganium eurycarpum* (p. 166)

            115. Heads not ball-shaped or roundish; leaves not triangular in cross-section
                116. Leaves joined at bases; flowers in terminal flat-topped inflorescences — Boneset, *Eupatorium perfoliatum* (p. 239)

                116. Leaves not joined at bases, sharply toothed with callous tips — Fireweed, *Erechtites hieracifolia* (p. 154)

    112. Flowers not in heads
        117. Flowers in ball-shaped clusters; tidal fresh marsh plants
            118. Flower clusters borne along a branched inflorescence; linear leaves often somewhat triangular in cross-section — Great Bur-reed, *Sparganium eurycarpum* (p. 166)

118. Flower clusters borne in leaf axils along stem, not in separate branched inflorescence; leaves not triangular in cross-section
    119. Leaves and stem having strong minty odor when crushed   Wild Mint, *Mentha arvensis* (p. 226)

    119. Leaves and stem lacking strong minty odor   Water Horehound, *Lycopus virginicus* (p. 225)

117. Flowers not in ball-shaped clusters
    120. Flowers less than ½ inch wide, usually much smaller
        121. Leaves oppositely arranged; tubular flowers with four to seven petallike lobes, flowers also pale purple; salt and brackish marsh plant from southern New Jersey south   Salt Marsh Loosestrife, *Lythrum lineare* (p. 216)

        121. Leaves not oppositely arranged
            122. Leaves both basal and alternately arranged; flowers five-lobed and bell-shaped   Water Pimpernel, *Samolus parviflorus* (p. 143)

            122. Leaves arranged in whorls, usually five or six per whorl; stem four-angled and rough with minute sharp teeth; flowers three-lobed   Dye Bedstraw, *Galium tinctorium* (p. 231)

    120. Flowers larger than ½ inch wide usually ¾–6½ inches
        123. Eight to twelve petals   Perennial Marsh Pink, *Sabatia dodecandra* (p. 145)

        123. Five petals
            124. Flowers larger than 4 inches wide, with or without red or purple centers   Rose Mallow, *Hibiscus moscheutos* (p. 138)

            124. Flowers ¾–1½ inches wide, with yellow center   Annual Marsh Pink, *Sabatia stellaris* (p. 146)

99. Flowers not white
    125. Red flowers arranged in terminal spikelike
        inflorescence   Cardinal Flower,
                                        Lobelia cardinalis
                                        (p. 234)

125. Greenish flowers
    126. Leaves rough with three to five lobes   Giant Ragweed,
                                        Ambrosia trifida
                                        (p. 235)

    126. Leaves not three-to-five-lobed
        127. Stems jointed
            128. Stems spiny   Arrow-leaved
                                        Tearthumb,
                                        Polygonum sagittatum
                                        (p. 200)

            128. Stems not spiny
                129. Flowers borne on conspicuous,
                      usually leafless inflorescences;
                      tidal fresh marsh plants
                      130. Flowers borne in loose,
                          erect spikes   Water Smartweed,
                                        Polygonum punctatum
                                        (p. 199)

                      130. Flowers borne on long,
                          drooping stalks arranged
                          in whorls along an
                          inflorescence   Swamp Dock, *Rumex*
                                        verticillatus (p. 201)

                129. Flowers borne in leaf axils, not
                      on conspicuous inflorescences;
                      plant along upper edges of salt
                      marshes   Bushy Knotweed,
                                        Polygonum
                                        ramosissimum (p. 126)

        127. Stems not jointed
            131. Flowers in ball-like clusters   Great Bur-reed,
                                          Sparganium
                                        eurycarpum (p. 166)

            131. Flowers on spikes from leaf axils
                132. Plant taller than 3½ feet; leaves
                      narrowly lance-shaped with
                      smooth margins; flowering
                      spikes axillary and terminal   Water Hemp,
                                        Amaranthus
                                        cannabinus (p. 132)

132. Plant less than 3 ½ feet tall; leaves broadly lance-shaped with coarse-toothed margins; flowering spikes axillary only

False Nettle,
Boehmeria cylindrica
(p. 197)

Wetland Plant Descriptions and Illustrations

The following descriptions and illustrations of common tidal wetland plants are general and intended to give the reader more characteristics to confirm that the plant in hand is the illustrated species, including life form, leaf type and arrangement, flower type and arrangement, fruit type, flowering period, habitat, and range. In addition, similar species, that is, other plants that may be confused with or closely related to the subject plant, are listed along with clues on how to distinguish among them. Diagnostic characteristics of these species may be illustrated as an inset. More technical plant descriptions can be found in several references including Fernald (1970), Gleason (1952), and Gleason and Cronquist (1963).

Scientific and common names for each described plant and its family are given. Scientific names follow the *National List of Scientific Plant Names* published by the U.S. Department of Agriculture Soil Conservation Service in 1982. Where the scientific name was recently changed, its previous name or synonym is indicated in parentheses below the current name. In the following descriptions, scientific names are usually represented by two Latin names and one or more abbreviations. For example, in the name *Typha angustifolia* L., *Typha* is the genus name, *angustifolia* the species name, and L. the abbreviated name of the author who first used this scientific name (in this case, Carolus Linnaeus). In general discussions, we usually drop the author's abbreviated name and refer to the plant as *Typha angustifolia* or by its common name, that is, narrow-leaved cattail.

All measurements are English units (i.e., inches and feet), because most readers are more familiar with them than with metric units. A scale is provided on the last printed page to aid in measurements. For those interested in learning the metric system, a conversion table is also presented there.

Plants are arranged according to their primary coastal wetland habitat (i.e., coastal waters, salt and brackish marshes, and tidal fresh marshes) and, within these groups, according to family. Because salt and brackish marshes have many species in common, plants characteristic of these habitats have been combined into a single section for organization purposes. The reader should also keep in mind that freshwater plants

may overlap into slightly brackish
marshes and that some species can
also be found along the landward
edges of salt marshes where
freshwater runoff or seepage is
significant.

Plants of Coastal Waters

Curly Pondweed

Potamogeton crispus L.

Pondweed Family
Potamogetonaceae
(Najadaceae or
Zosteraceae)

Description. Rooted, submerged
aquatic plant; stem with few
branches; simple, wavy-margined,
sessile submerged leaves (1 ¼–
3 ¼ inches long and ¼–½ inch
wide) with three to five nerves and
floating leaves absent; flowers borne
in dense spikes on end of a stalk
(peduncle, 1–2 inches long); fruit
nutlet (achene).

Flowering period. May through
September.

Habitat. Brackish and tidal fresh
waters (introduced); native of
Europe.

Range. Massachusetts to
Minnesota, south to Virginia and
Missouri.

Similar species. Other *Potamogeton*
spp. (Pondweeds) do not have curly,
wavy-margined leaves.

× ¾

Ribbonleaf Pondweed

Potamogeton epihydrus Raf.

Pondweed Family
Potamogetonaceae
(Najadaceae or
Zosteraceae)

Description. Rooted, submerged, and floating-leaved aquatic plant; stems flattened, often branched; two types of leaves—(1) simple, entire, linear submerged leaves (up to 8 inches long) with five to seven nerves and lacking leaf stalk (petiole), and (2) simple, entire, egg-shaped to spoon-shaped (spatulate) floating leaves (1¼–2⅘ inches long and ⅓–⅘ inch wide) with eleven to twenty-seven nerves and with leaf stalk (petiole); flowers borne in numerous dense spikes (⅖–1⅕ inches long) on end of stalk (peduncle, 1–2 inches long); fruit nutlet (achene).

Flowering period. July through September.

Habitat. Tidal fresh waters; ponds and slow-moving streams.

Range. Newfoundland and Quebec to southern Alaska, south to Georgia, Iowa, Colorado, and California.

Similar species. Numerous pondweeds are present in tidal fresh waters. *Potamogeton crispus* (Curly Pondweed) has curly, wavy-margined leaves. *P. amplifolius* (Largeleaf Pondweed) has lance-shaped to broadly lance-shaped, somewhat folded submerged leaves with twenty-five to fifty nerves. *P. pulcher* (Heartleaf Pondweed) has floating leaves with heart-shaped or rounded bases and black-spotted stems and leaf stalks. *P. nodosus* (Longleaf Pondweed) has linear to narrowly lance-shaped submerged leaves with seven to fifteen nerves.

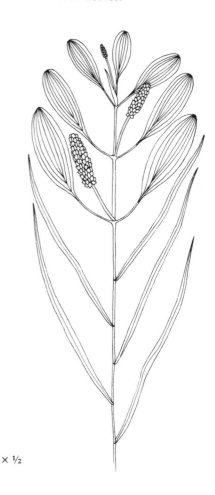

× ½

Sago Pondweed

Potamogeton pectinatus L.

Pondweed Family
Potamogetonaceae
(Najadaceae or
Zosteraceae)

Description. Rooted, submerged
aquatic plant; stems much branched;
simple, entire, linear leaves (1−4
inches long), threadlike, tapering to
a long point, with one vein; several
whorls of minute flowers borne on
spikes (½−1½ inches long) on
stalks (peduncles, to 4 inches long);
fruit nutlet (achene).

Flowering period. Summer.

Fruiting period. June to September.

Habitat. Brackish and tidal fresh
waters; shallow fresh (calcareous)
waters of lakes and slow-flowing
streams.

Range. Quebec and Newfoundland
to Alaska and British Columbia,
south to Florida, Texas, and southern
California.

Similar species. Other *Potamogeton*
with only linear submerged leaves
are *P. foliosus* (Leafy Pondweed),
P. pusillus (Baby Pondweed), *P.
zosteriformis* (Flat-stem Pondweed),
and *P. robbinsii* (Robbins'
Pondweed). Their leaves all have
more than three veins: three to five
in *P. foliosus* and *P. pusillus*, usually
more than twenty-five and up to
thirty-five in *P. zosteriformis*, and
twenty to sixty in *P. robbinsii*.
Ruppia maritima (Widgeon Grass)
has two-flowered spikes enclosed in
a sheathing leaf base before

maturity, whereas *P. pectinatus*
usually has more than two flowers
arranged in whorls along a spike.

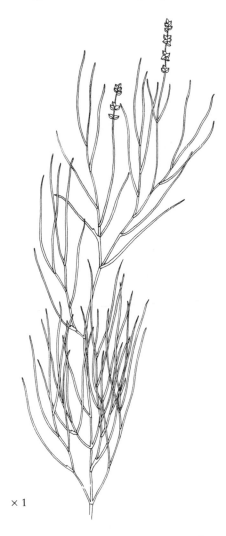

× 1

85

Clasping-leaved Pondweed or Redhead-grass

Potamogeton perfoliatus L.

Pondweed Family
Potamogetonaceae
(Najadaceae or
Zosteraceae)

Description. Rooted, submerged aquatic plant; stems slender, usually short, and much branched with internodes ½–1¼ inches long; egg-shaped or rounded leaves (½–3¼ inches long, usually ½–1½ inches long) with round or broad tips, base often heart-shaped and clasping stem, with three, sometimes five, prominent nerves and several weaker ones; flowers borne in dense spikes on end of stalk (peduncle, 1¼–4¾ inches long); fruit nutlet (achene).

Flowering period. Summer.

Fruiting period. July to October.

Habitat. Brackish and tidal fresh waters; ponds and slow-moving streams.

Range. Newfoundland and Quebec to Ohio, south to Florida and Louisiana.

Similar species. Potamogeton richardsonii has lance-shaped, clasping submerged leaves (2–4 inches long). *P. crispus* (Curly Pondweed) has wavy margined submerged leaves. Other *Potamogeton* in tidal fresh waters with only submerged leaves have linear leaves (*P. foliosus*, *P. pectinatus*, *P. pusillus*, and *P.*

zosteriformis). Other *Potamogeton* have two types of leaves—floating and submerged.

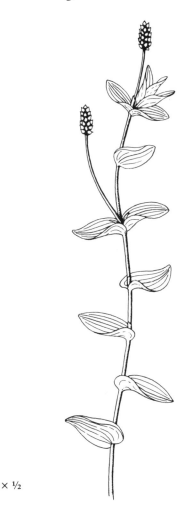

× ½

Widgeon Grass

Ruppia maritima L.

Pondweed Family
Potamogetonaceae
(Najadaceae or
Zosteraceae)

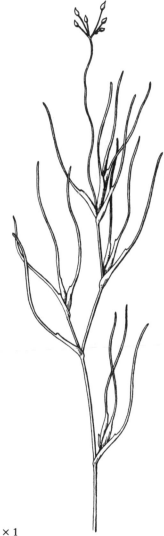

Description. Rooted, submerged aquatic plant; stems simple or branched and up to 3 feet long; simple, entire, linear leaves (1¼–4 inches long), threadlike, with leaf sheaths present, alternately arranged; flowers and fruits borne on stalks (⅕–12 inches long); fruit fleshy (drupe).

Flowering period. Summer.

Fruiting period. July to October.

Habitat. Saline and brackish waters, rarely tidal fresh waters, salt ponds and pools within salt marshes; inland saline waters, rarely fresh waters.

Range. Newfoundland to Florida and Mexico; along Pacific coast from Washington to California; inland from western New York to British Columbia.

Similar species. The leaves of *Zannichellia palustris* (Horned Pondweed) are also threadlike but oppositely arranged; its flowers and fruits are very short-stalked.

× 1

Horned Pondweed

Zannichellia palustris L.

Pondweed Family
Potamogetonaceae
(Najadaceae or
Zosteraceae)

Description. Rooted, submerged aquatic plant; stems very slender, fragile, and branched; simple, entire, linear leaves (up to 4 inches long), threadlike, oppositely arranged; minute flowers borne in leaf axils enclosed by a sheath; fruit oblong nutlet (achene).

Flowering period. July to October.

Habitat. Brackish and tidal fresh waters; inland fresh and alkaline waters.

Range. Newfoundland and Quebec to Alaska, south to Florida, Texas, and Mexico.

Similar species. The leaves of *Najas guadalupensis* (Naiad or Bushy Pondweed) are also linear and oppositely arranged but are shorter, more crowded, very finely toothed (microscopically), and not threadlike. The leaves of *Ruppia maritima* (Widgeon Grass) are also threadlike but are alternately arranged.

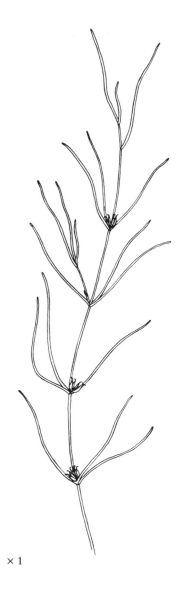

× 1

88

Eelgrass

Zostera marina L.

Pondweed Family
Potamogetonaceae
(Najadaceae or
Zosteraceae)

Description. Rooted, submerged
aquatic plant, sometimes exposed at
extreme low tides; stems slender
and branched; simple, entire, linear,
ribbonlike leaves (up to 2 feet long
and ½ inch wide) with three to five
distinct nerves; inconspicuous
(hidden) flowers borne on one side
of leaf enclosed within a sheath;
fruit cylinder-shaped seed.

Flowering period. Summer.

Habitat. Shallow estuarine saline
waters in sheltered bays and coves,
occasionally tidal flats.

Range. Greenland and Labrador to
Florida; also along the Pacific coast.

Similar species. Leaves of *Vallisneria
americana* (Wild Celery or Tape
Grass) look somewhat similar, but
this plant grows in slightly brackish
and tidal fresh coastal waters.

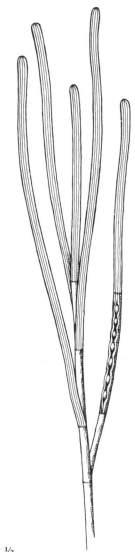

× ½

Naiad or Bushy Pondweed

Najas guadalupensis (Spreng.)
Morong.

Naiad Family
Najadaceae

Description. Rooted, submerged
aquatic plant; stems very long and
leafy; simple, somewhat entire
(actually with twenty or more
microscopic teeth along each
margin), linear leaves (less than
1 inch long), dark green or olive-
colored, expanded at base and
sloping gradually above to a
rounded or fine-pointed end,
oppositely arranged; small flowers
and fruits borne singly in leaf axils;
fruit purplish brown enclosing
straw-colored ribbed seed.

Flowering period. August to
October.

Habitat. Tidal fresh waters; inland
lakes and ponds.

Range. Southeastern Massachusetts
and New York to South Dakota and
Idaho, south to Florida and Texas;
also in the Pacific states.

Similar species. Najas flexilis
(Nodding Water Nymph) has similar
leaf bases, but its leaves are pale
green and tapering to a long curved
tip. Other *Najas* in tidal waters have
greatly expanded leaf bases and
toothed leaf margins. *N. gracillima*
and *N. minor* have six to twenty
and six to fifteen spiny teeth (visible
through 10-power magnification)
per margin, respectively. Of these
two, *N. gracillima* is more
widespread, whereas *N. minor* is

locally common in the Hudson
River. The leaves of *Zannichellia
palustris* (Horned Pondweed) are
also linear and oppositely arranged
but are usually narrower
(threadlike), longer (more than
1 inch long), less crowded, and
entire (without teeth).

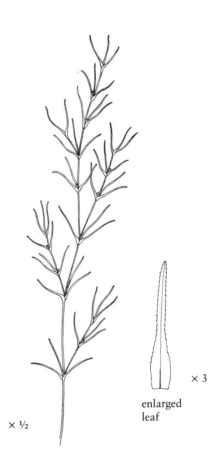

× ½

× 3

enlarged
leaf

Waterweed

Elodea canadensis Michx.
(*Anacharis canadensis* [Michx.] Rich.)

Frog's-bit Family
Hydrocharitaceae

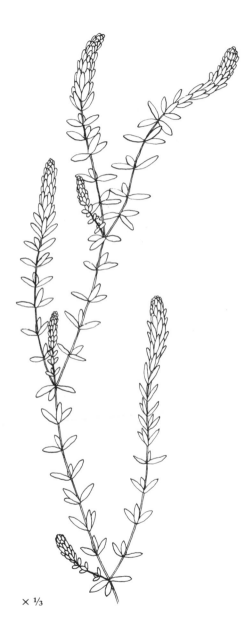

Description. Rooted, submerged aquatic plant, sometimes floating at surface in shallow water; stems many-branched, often forming dense masses; elongate, sessile, dark green leaves usually twice as long as wide (about ⅗ inch long and ⅕ inch wide), mostly drooping downward, with finely toothed margins, mostly arranged in whorls of threes; male and female flowers borne on stalks (pedicels) arising from tubular structure (spathe) in leaf axil, male stalks longer than female stalks; fruit cylinder-shaped capsule.

Flowering period. July to September.

Habitat. Tidal fresh waters; inland waters, often calcareous.

Range. Quebec to Saskatchewan and Washington, south to North Carolina, Alabama, Oklahoma, and California.

Similar species. Elodea nuttalli has narrower leaves (less than ⅕ inch wide), and its male flowers are not borne on a long stalk. *Egeria densa* (South American Elodea, formerly *Elodea densa*) has longer leaves (⅘ – 1⅖ inches) arranged in whorls of fours to sixes.

× ⅓

Wild Celery or Tape Grass
Vallisneria americana Michx.

Frog's-bit Family
Hydrocharitaceae

Description. Rooted, submerged aquatic plant; stems buried in mud; simple, entire, very thin linear basal leaves (up to 7 feet long), ribbonlike; two types of flowers—male flowers borne in structure (spathe) at base of leaves and released to water's surface, female flowers borne on long stalk (peduncle) reaching water's surface; fruit cylinder-shaped capsule, peduncle coils after fertilization pulling fruit beneath water.

Flowering period. July to October.

Habitat. Tidal fresh waters, occasionally slightly brackish waters; inland waters.

Range. New Brunswick, Nova Scotia, and Quebec to North Dakota, south to Florida and Texas.

Similar species. Leaves of *Zostera marina* (Eelgrass) look somewhat similar, but this plant grows only in saline coastal waters.

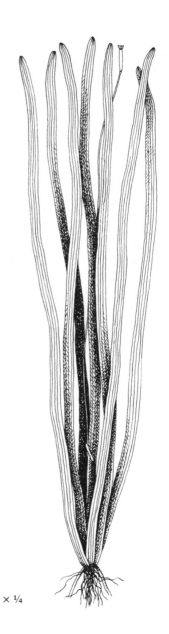

× ¼

92

Big Duckweed

Spirodela polyrhiza (L.) Schleid.

Duckweed Family
Lemnaceae

Description. Free-floating, surface-water aquatic plant; stem lacking; leaflike structure (thallus) broadly oval-shaped (¹⁄₁₀–²⁄₅ inch long), dark green above, purple below, with six to eighteen, usually seven, nerves and six to eighteen rootlets; flowers in pouches (rarely seen).

Flowering period. Summer.

Habitat. Tidal fresh waters; freshwater lakes, ponds, and slow-flowing streams.

Range. Nova Scotia to British Columbia, south to Florida, Texas, and Mexico.

Similar species. Lemna minor (Little Duckweed) has only one rootlet per plant (thallus).

top view

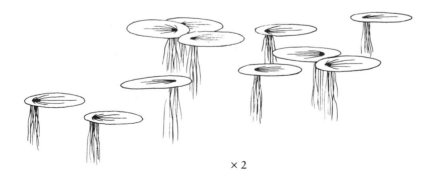

× 2

93

White Water Lily or Fragrant Water Lily

Nymphaea odorata Soland. in Ait.

Water Lily Family
Nymphaeaceae

Description. Rooted, floating-leaved aquatic plant; perennial from elongate, branched rhizome; roundish floating leaves (up to 10 inches wide) notched at base, green above and normally purplish below, attached to rhizome by long purple to red stalk (petiole); large, showy, fragrant white flower (2–6 inches wide) with many petals (seventeen to thirty-two) borne singly on a long stalk.

Flowering period. June into September.

Habitat. Tidal fresh waters; inland shallow waters of lakes and ponds.

Range. Newfoundland to Manitoba and northern Minnesota, south to Florida and Louisiana.

Similar species. Nuphar luteum (Spatterdock) has erect, fleshy heart-shaped leaves with a distinct midrib below, and its flower is yellow with five or six petals. *N. luteum* ssp. *variegatum* (Bullhead Lily) is like Spatterdock but has floating, fleshy, heart-shaped leaves.

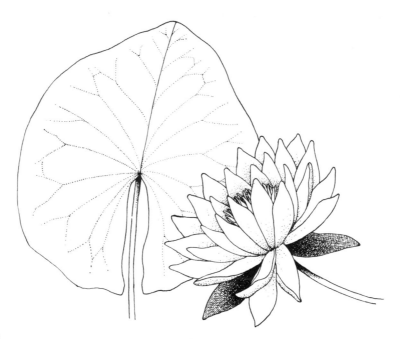

× ⅓

94

Coontail

Ceratophyllum demersum L.

Hornwort Family
Ceratophyllaceae

Description. Free-floating, submerged aquatic plant; stems much branched and forming large masses; compound, toothed linear leaves (²⁄₅–1 ⅕ inches long), two or three times divided, arranged in five to twelve whorls; minute flowers borne singly in leaf axils; fruit nutlet (achene).

Flowering period. July through September.

Habitat. Tidal fresh waters; inland lakes and slow-flowing streams.

Range. Quebec to northern British Columbia, south to Florida, Texas, and California.

× ½

Plants of Salt and Brackish Marshes

Narrow-leaved Cattail
Typha angustifolia L.

Cattail Family
Typhaceae

Description. Medium-height to tall, erect herbaceous plant, up to 6 feet high; perennial; simple, entire, elongate, linear basal leaves (⅕–½ inch wide), flattened (plano-convex), sheathing at base and ascending along stem in an apparent alternately arranged fashion, usually less than ten leaves; inconspicuous flowers borne on long stalk and arranged in two terminal cylinder-shaped spikes (male spike above female spike) separated by a space, female spike green in spring and brown in summer at maturity and persistent in winter, male spike covered with yellow pollen grains at maturity and then disintegrating (nonpersistent).

Flowering period. Late May through July.

Habitat. Brackish and tidal fresh marshes (regularly and irregularly flooded zones); inland fresh and alkaline marshes.

Range. Nova Scotia, Quebec, and Ontario, south to Florida and Texas, especially abundant along the coast.

Similar species. Typha latifolia (Broad-leaved Cattail) grows taller (to 9 feet) and has wider leaves (to 1 inch) and no space between male and female spikes. *T. domingensis* (Southern Cattail) occurs from Delaware and Maryland south; it is

much taller (8–13 feet), with ten or more leaves and a space between male and female spikes.

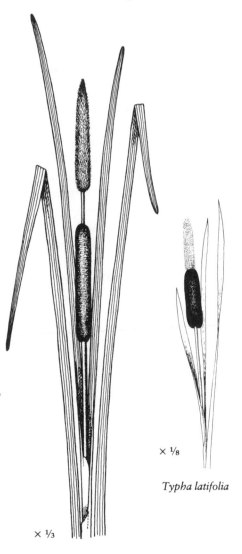

× ⅛

Typha latifolia

× ⅓

99

Seaside Arrow Grass

Triglochin maritimum L.

Arrow Grass Family
Juncaginaceae

Description. Low to medium-height, erect, fleshy herbaceous plant, 8–32 inches tall; perennial; basal, linear, fleshy leaves (up to 20 inches long) with conspicuous sheaths; numerous small greenish flowers borne terminally on flowering spike (scape); fruit cylinder-shaped with twelve narrow wings.

Flowering period. May through August.

Habitat. Irregularly flooded salt and brackish marshes and shores; inland marshes and shores.

Range. Labrador to Alaska, south to New Jersey and Delaware (rarely to Maryland); inland to Newfoundland and North Dakota, south to Arizona and New Mexico.

Similar species. Triglochin palustre (Marsh Arrow Grass) also occurs in brackish marshes in our area; its stems are very slender, and its fruits are thin capsules with club-shaped ends. The leaves of *Plantago maritima* (Seaside Plantain) are also fleshy, linear, and basal but lack leaf sheaths.

× ½

Stiff-leaf Quackgrass

Agropyron pungens (Pers.) Roem. &
J. A. Schultes

Grass Family
Gramineae (Poaceae)

Description. Medium-height, erect
grass, 1½–3½ feet tall; perennial
from long, slender, creeping
rhizomes; stem not hollow but pithy
at maturity, especially near top;
elongate, stiff, linear leaves, usually
rolled inwardly at distal ends, upper
surface rough with coarse ribs, gray-
greenish or bluish green in color;
inconspicuous flowers borne on
dense terminal spikes (2–6 inches
long), usually four-angled at joints,
spikelets with seven to eleven
flowers.

Flowering period. July into
September.

Habitat. Upper edges of salt
marshes, usually sandy areas.

Range. Nova Scotia to southeastern
Massachusetts.

× ½

Creeping Bent Grass

Agrostis stolonifera L.
var. *compacta* Hartm.
(*Agrostis palustris* Huds.)

Grass Family
Gramineae (Poaceae)

Description. Low to medium-height,
erect grass, generally 4–16 inches
tall, rarely to 32 inches; densely
matted perennial from creeping
stems (stolons) above or just below
marsh surface; stems hollow and
round, prostrate at base and
ascending; leaves (less than ⅕ inch
wide) sometimes rolled inwardly;
oblong (cylinder-shaped) terminal
inflorescence (panicle, to 7 inches
long) with appressed or closely
appressed branches of unequal
length and arranged in whorls.

Flowering period. June through
September.

Habitat. Irregularly flooded
brackish and tidal fresh marshes;
inland wet meadows and shores.

Range. Newfoundland, Labrador,
and Quebec to British Columbia,
south to Virginia, Indiana,
Minnesota, New Mexico, and
California.

× ½

Spike Grass or Salt Grass

Distichlis spicata (L.) Greene

Grass Family

Gramineae (Poaceae)

Description. Low-growing, erect grass, 8–16 inches tall; perennial from creeping rhizomes, often forming dense mats; stems stiff, hollow, and round; numerous linear leaves (2–4 inches long) with smooth margins usually rolled inwardly and sheaths overlapping, distinctly two-ranked; terminal inflorescence (panicle, to 2½ inches long) bears one of two types of crowded spikelets on separate plants (dioeceous), male spikelets with eight to twelve flowers, female spikelets usually five-flowered (four to nine).

Flowering period. August into October.

Habitat. Irregularly flooded salt marshes, often intermixed with *Spartina patens* (Salt Hay Grass) or in pure stands in wet depressions, and brackish marshes.

Range. New Brunswick and Prince Edward Island, south to Florida and Texas; locally inland to Missouri; also along the Pacific coast.

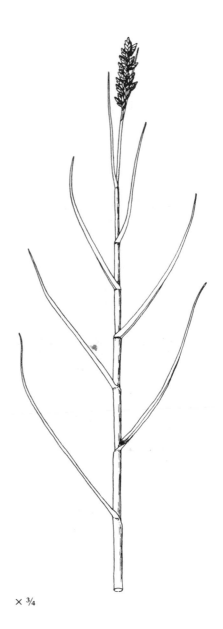

× ¾

Red Fescue

Festuca rubra L.

Grass Family
Gramineae (Poaceae)

Description. Low to medium-height, erect grass, 1−3½ feet tall; perennial in loose clumps from matted, short rhizomes; stems smooth, hollow, and round, often reclining at base and ascending; linear leaves, usually rolled inwardly, with reddish to brownish lower leaf sheaths quickly becoming loose fibers, upper sheaths smooth; inconspicuous flowers borne in narrow terminal inflorescence (panicle, 2−9 inches long) with ascending branches, spikelets bear three to ten flowers with bristlelike awns.

Flowering period. June to August.

Habitat. Irregularly flooded brackish marshes and upper edges of salt marshes; inland marshes, fields, and roadsides.

Range. Labrador and Quebec to Wisconsin, south to Virginia (along coast) and to Georgia (in mountains).

Similar species. Hierochloe odorata (Sweet Grass) also occurs in the upper salt marsh; it has elongate leaf sheaths, bladeless or short-bladed leaves (less than 1 ¼ inches long), and a wide, spreading panicle with often somewhat drooping branches.

× ⅓

Hierochloe odorata

× ¾

Switchgrass

Panicum virgatum L.

Grass Family

Gramineae (Poaceae)

Description. Medium-height to tall, erect grass, up to 6½ feet high, forming dense clumps; perennial from hard, scaly rhizomes; stems stout, round, and erect; smooth, long, tapered leaves (up to 20 inches long and ⅘ inch wide), sometimes with few hairs at base; open terminal inflorescence (panicle, 8–16 inches long) many-branched and pyramid-shaped, branches fairly open with many spikelets on slender stalks.

Flowering period. July to September.

Habitat. Upper edges of salt marshes and irregularly flooded brackish and tidal fresh marshes; open woods, prairies, dunes, and shores.

Range. Nova Scotia and Quebec to Manitoba and Montana, south to Florida, Texas, and Arizona.

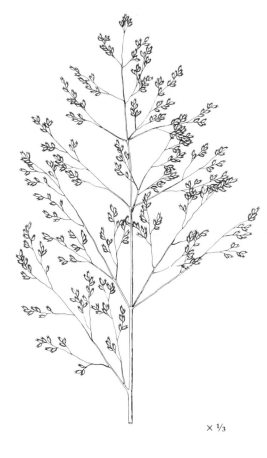

× ¹⁄₂₅

× ⅓

Common Reed

Phragmites australis (Cav.) Trin. ex
Steud. (*Phragmites communis* Trin.)

Grass Family
Gramineae (Poaceae)

Description. Tall, erect grass, 6½–
14 feet high, usually forming
dense stands; perennial from stout
rhizomes, sometimes creeping on
surface; stems round, hollow, and
erect; long, flat, tapering (long-
pointed) leaves (up to 24 inches
long and 2 inches wide), distinctly
arranged in two ranks; dense, many-
branched terminal inflorescence
(panicle, 8–16 inches long) with
silky, light brown hairs beneath on
stem, branches usually somewhat
drooping, flower clusters usually
purplish when young and white or
light brown and feathery when
mature.

Flowering period. Late July to
October.

Habitat. Brackish and tidal fresh
marshes (regularly and irregularly
flooded zones), also upper edges of
salt marshes and old spoil deposits;
inland marshes, swamps, wet shores,
ditches, and disturbed areas.

Range. Nova Scotia and Quebec to
British Columbia, south to Florida,
Texas, and California.

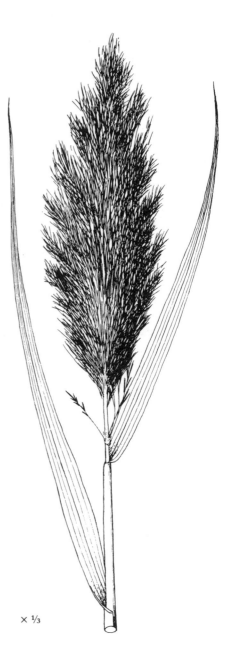

× ⅓

Seashore Alkali Grass or Goose Grass

Puccinellia maritima (Huds.) Parlat.

Grass Family
Gramineae (Poaceae)

Description. Low to medium-height, erect grass, 8–32 inches tall; perennial growing in clumps; stems hollow and round; leaves (less than ⅕ inch wide) often rolled inwardly, arranged in two ranks; terminal inflorescence (panicle, 2–8 inches long) narrow with ascending branches bearing spikelets with four to eleven flowers.

Flowering period. June into September.

Habitat. Irregularly flooded salt marshes and shores.

Range. Nova Scotia to Rhode Island, locally to southern Pennsylvania.

Similar species. Other *Puccinellia* usually have less flowers per spikelet than *P. maritima*. *P. fasciculata* (Torrey Alkali Grass) and *P. pumila* (Arctic Alkali Grass) have two to five flowers per spikelet, whereas *P. distans* has four to six flowers per spikelet. *P. fasciculata* has a compact inflorescence with the lowest branches usually covered with flowers for their entire length, except near the base, and has stems often slightly bent at the nodes. *P. pumila* has an inflorescence with erect upper branches and often spreading lower branches. *P. distans* has an open inflorescence with spreading branches usually naked for more than half their length, bearing spikelets at the distal end, and with lower branches often arranged in clusters of fours or fives; its stem is prostrate at the base and ascending elsewhere.

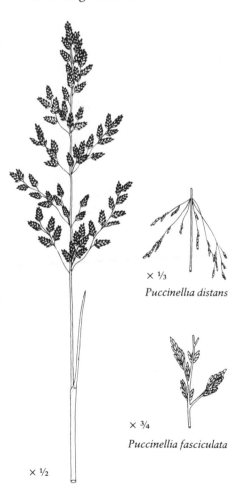

× ½

× ⅓
Puccinellia distans

× ¾
Puccinellia fasciculata

Foxtail Grass

Setaria geniculata (Lam.) Beauv.

Grass Family
Gramineae (Poaceae)

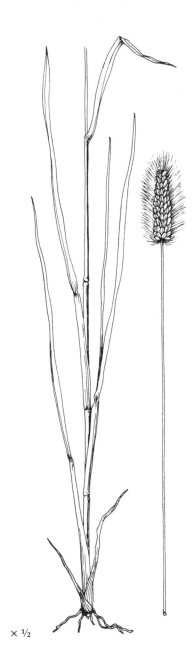

Description. Medium-height, erect grass, 1½–2½ feet tall; perennial from short, knotty rhizomes; stems round, hollow, and erect or sometimes lying flat on ground at base, then ascending; long, tapering, mostly flat leaves (up to 8 inches long and ¼ inch wide); dense terminal spikelike inflorescence (panicle, 1–4 inches long) with many spikelets and light brown bristles (five or more below each spikelet).

Flowering period. July to October.

Habitat. Upper edges of salt marshes; moist to dry ground and waste places.

Range. Massachusetts, south to Florida and Texas; inland in the North to Pennsylvania, Illinois, Kansas, New Mexico, and California.

Similar species. Setaria magna (Giant Foxtail) occurs in brackish marshes from New Jersey south; it grows to 13 feet tall and has one to three bristles below each spikelet. *Echinochloa walteri* (Walter Millet) also has bristled spikelets, but its panicle is branched and not a single dense spike.

× ½

108

Smooth Cordgrass or Saltwater Cordgrass

Spartina alterniflora Loiseleur

Grass Family
Gramineae (Poaceae)

Description. Low to tall, erect grass, 1–8 feet high; perennial; stems stout, round, and hollow, soft and spongy at base; elongate smooth leaves (up to 16 inches long and ½ inch wide) tapering to a long point with inwardly rolled tip, leaf margins smooth or weakly rough, sheath margins hairy; narrow terminal inflorescence (panicle, 4–12 inches long) composed of five to thirty spikes (2–4 inches long) alternately arranged and appressed to main axis with ten to fifty sessile spikelets each.

Flowering period. July through September.

Habitat. Salt and brackish marshes (regularly and irregularly flooded zones).

Range. Quebec and Newfoundland to Florida and Texas.

Similar species. Other *Spartina* members have more open panicles. *S. alterniflora* usually is the only plant in the regularly flooded low marsh zone of salt and brackish marshes. Salt crystals can often be seen on its leaves during the growing season. Two growth forms are generally recognized: (1) short form (less than 1½ feet, often having yellowish green leaves, and characteristic of irregularly flooded high marsh) and (2) tall form (greater than 1½ feet and typical of regularly flooded low marsh).

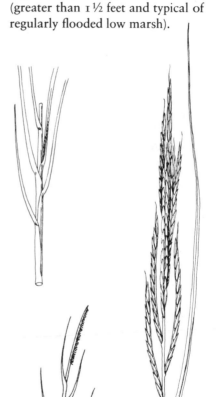

× 1/3

× 1/16

habit

Big Cordgrass

Spartina cynosuroides (L.) Roth

Grass Family
Gramineae (Poaceae)

Description. Tall, erect grass, 3½–10 feet high; perennial; stems stout, round, and hollow; elongate leaves (up to 28 inches long and 1 inch wide) tapering to a point, margins very rough; open terminal inflorescence (panicle, 4–12 inches long) composed of twenty to fifty erect, crowded spikes (1¼–2⅘ inches long and uppermost spikes usually shorter than lower ones), each with many (up to seventy) densely overlapping spikelets.

Flowering period. August into October.

Habitat. Irregularly flooded salt, brackish, and tidal fresh marshes.

Range. Massachusetts to Florida and Texas.

Similar species. Spartina pectinata (Slough Grass or Prairie Cordgrass) looks very similar, but it has bristle-tipped spikelets and usually less than twenty spikes per panicle.

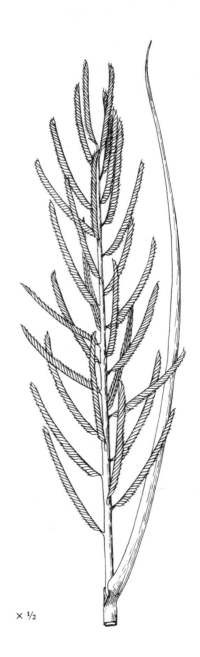

× ½

Salt Hay Grass or
Salt Meadow Grass
Spartina patens (Ait.) Muhl.

Grass Family
Gramineae (Poaceae)

Description. Low to medium-height, erect or spreading grass, 1–3 feet tall, often forming cowlicked mats; perennial with long rhizomes; stems slender (wirelike), stiff, and hollow; very narrow linear leaves (less than ⅕ inch wide and up to 1½ feet long) with margins rolled inwardly; open terminal inflorescence (panicle, up to 8 inches long) usually composed of three to six spikes (⅘–2 inches long) alternately arranged and diverging from main axis at 45–60-degree angles, each with twenty to fifty densely overlapping spikelets (⅕–½ inch long).

Flowering period. Late June into October.

Habitat. Irregularly flooded salt and brackish marshes, often forming cowlicked mats, and reported to occur at times in regularly flooded zone; (var. *monogyna*—on wet beaches, sand dunes, and borders of salt marshes); also inland saline areas.

Range. Quebec to Florida and Texas; inland in New York and Michigan.

Similar species. Other *Spartina* members have stout stems, not wirelike. *S. patens* grows in two forms; the typical form lies flat (decumbent) with upward-spreading stems that create the "cow-licks" of the high salt marsh, whereas the variety *monogyna* grows upright and straight.

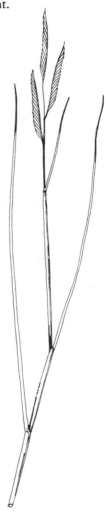

× ½

III

Slough Grass or Prairie Cordgrass

Spartina pectinata Link

Grass Family
Gramineae (Poaceae)

Description. Medium-height to tall, erect grass, 3–6½ feet high; perennial; stems stout, round, and hollow; very long leaves (up to 4 feet long and ⅗ inch wide) tapering to a long, threadlike tip, very rough margins rolled inwardly when dry; open terminal inflorescence (panicle, 4–16 inches long) usually composed of ten to twenty short-stalked, erect spikes (2–4 inches), each with many densely overlapping, bristle-tipped spikelets (½ inch long).

Flowering period. July into September.

Habitat. Irregularly flooded brackish and tidal fresh marshes, occasionally upper borders of salt marshes; inland marshes, shores, and wet prairies.

Range. Newfoundland and Quebec to Alberta and Washington, south to New Jersey, western North Carolina, Arkansas, and Texas.

Similar species. Spartina cynosuroides (Big Cordgrass) looks very similar, but it does not have bristle-tipped spikelets and usually has more than thirty spikes per panicle.

× ⅓

Marsh Straw Sedge

Carex hormathodes Fernald

Sedge Family
Cyperaceae

Description. Low to medium-height, erect, grasslike herbaceous plant, 8–38 inches tall; perennial in dense clumps; stems slender, sharply three-angled, smooth (except at top), erect, and spreading; simple, entire, elongate, narrow linear leaves (less than 1/10 inch wide), smooth except rough near tip; inconspicuous flowers borne in dense rust to straw-colored clusters (heads), often subtended by a bristlelike bract, usually three to nine (sometimes up to fifteen) clusters forming a terminal inflorescence (up to 3 inches long), clusters somewhat egg-shaped and gently tapered at both ends (up to 3/5 inch long); fruit nutlet (achene) enclosed in a flattened sac (perigynium).

Flowering period. Late May through August.

Habitat. Irregularly flooded brackish and tidal fresh marshes; also sandy areas and rocks near the coast.

Range. Newfoundland and Quebec, south to Virginia.

Similar species. Carex scoparia also occurs in brackish and tidal fresh marshes; its perigynia are covered by scales lacking bristlelike (awn) tips, whereas the scales of *C. hormathodes* are awn-tipped. Identification of sedges is difficult and requires examination of achenes and perigynia. In tidal fresh marshes, numerous sedges may be present including *C. alata, C. albolutescens, C. annectens, C. comosa, C. crinita, C. lacustris, C. lupulina, C. lurida, C. normalis, C. rostrata, C. stricta,* and *C. tremuloides.* Of these, *C. stricta* (Tussock Sedge) is most common; it forms distinct clumps called tussocks. Other sedges should be identified by using a taxonomic reference, such as Fernald (1970) or Gleason (1952).

Scale

× 3

× 2/3

113

Salt Marsh Sedge

Carex paleacea Schreb. ex Wahlenb.

Sedge Family
Cyperaceae

Description. Medium-height, erect, grasslike herbaceous plant, 1–3 feet tall, growing solitary or in loose clumps; perennial; elongate, somewhat erect, and slightly drooping linear leaves (1/10–1/3 inch wide) tapering to a long, thin point, arranged in three ranks; inconspicuous flowers borne in two types of spikes (male and female), male spikes terminal, one to four in number on drooping, slender stalks (peduncles), and female spikes (4/5–3 inches long), two to six in number on widely spreading or drooping peduncles, flowers covered by brown scales with a pale midvein; fruit seedlike covered by a firm sac (perigynium) having a long, pointed beak.

Flowering period. June through August.

Habitat. Upper edges of salt marshes and irregularly flooded brackish marshes.

Range. Greenland, Labrador, and Quebec, south to Massachusetts.

Similar species. Carex salina also occurs in salt marshes from Massachusetts north; its female spikes are erect and not wide-spreading or drooping. *C. hormathodes* occurs in brackish and tidal fresh marshes; its erect, terminal inflorescence is composed of three to nine, sometimes fifteen, clusters of flowering heads.

× 1/4

Nuttall's Cyperus
Cyperus filicinus Vahl

Sedge Family
Cyperaceac

Description. Low to medium-height, erect, grasslike herbaceous plant, 4–16 inches tall; annual; stems three-angled; elongate, linear leaves (usually less than $\frac{1}{10}$ inch wide); inconspicuous flowers borne in sessile or stalked spikes (up to 5 inches long) forming clusters, several clusters forming terminal inflorescence (umbel) subtended by several leafy bracts (up to 10 inches long), flowers covered by green (immature) or straw-colored (mature) sharp-pointed scales with three to five nerves clustered together at center; fruit flattened nutlet (achene).

Flowering period. August into October.

Habitat. Brackish marshes (regularly and irregularly flooded zones); sandy coastal beaches and, rarely, inland shores of ponds.

Range. Southern Maine to Florida and Louisiana.

Similar species. Cyperus filiculmis (Slender Cyperus) occurs in brackish and tidal fresh marshes; it is a perennial plant with bulblike rhizomes, it has somewhat rough-margined leaves, and its scales have seven to thirteen well-spaced nerves. Other *Cyperus* spp. resembling *C. filicinus* are found in tidal fresh marshes. *C. flavescens* (Yellow Cyperus) has yellowish green scales about half as wide as long, whereas the scales of *C. filicinus* are roughly three times as long as wide. *C. rivularis* (Shining Cyperus) has reddish brown, blunt-tipped scales at maturity.

× 2

scale

× ½

Dwarf Spike-rush

Eleocharis parvula
(Roem. & J. A. Schultes)
Link ex Bluff & Fingerh.

Sedge Family
Cyperaceae

Description. Low-growing, erect, grasslike herbaceous plant, usually less than 3 inches but up to 5 inches tall, forming mats; stems spongy and threadlike with no apparent leaves, leaves actually reduced to stem sheaths; inconspicuous flowers covered by green, straw-colored, or brown scales borne on a single terminal budlike spikelet; fruit three-angled nutlet (achene).

Flowering period. July to October.

Habitat. Salt and brackish marshes (regularly and irregularly flooded zones); wet inland saline soils.

Range. Newfoundland to Florida and Texas; inland locally in western New York, Michigan, and Missouri; also on Pacific coast.

Similar species. Other spike-rushes of coastal marshes are generally taller and not mat-forming, especially *Eleocharis halophila,* *E. obtusa, E. rostellata,* and *E. palustris.* Positive identification requires examining the fruits (achenes) and reference to a taxonomic manual.

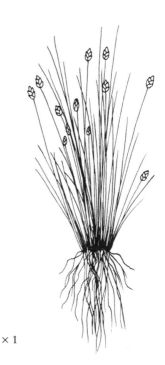

× 1

Beaked Spike-rush

Eleocharis rostellata (Torr.) Torr.

Sedge Family
Cyperaceae

Description. Medium-height to tall, erect grasslike plant, 1–3½ feet high; perennial in dense clumps; stems elongate and flattened with no apparent leaves, leaves actually reduced to stem sheaths; inconspicuous flowers covered by brownish scales borne on a single terminal budlike spikelet; fruit three-angled nutlet (achene).

Flowering period. July into October.

Habitat. Irregularly flooded salt, brackish, and tidal fresh marshes; calcareous inland marshes and swamps.

Range. Nova Scotia to Florida, along the coast; inland locally from New York to Wisconsin, south to Ohio and Indiana; also in western states.

Similar species. A few other tall spike-rushes commonly occur in coastal wetlands: *Eleocharis palustris* (Common Spike-rush), *E. obtusa* (Blunt Spike-rush), and *E. halophila* (Salt Marsh Spike-rush). They all have lens-shaped nutlets. To distinguish among them requires close examination of nutlets. Another spike-rush, *E. quadrangulata* (Square-stemmed Spike-rush), is present in tidal fresh marshes and shallow water; it has distinct four-angled stems, and its spikelet is barely thicker than its stem. Other spike-rushes reported from northeastern coastal wetlands include *E. albida* (Maryland south), *E. ambigens*, *E. olivacea*, and *E. smallii*.

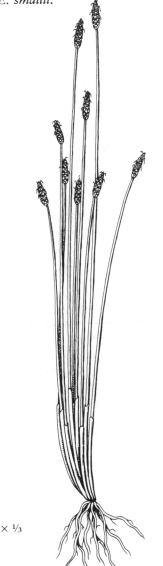

× ⅓

117

Salt Marsh Fimbristylis

Fimbristylis castanea (Michx.) Vahl

Sedge Family
Cyperaceae

Description. Low to medium-height, erect, grasslike herbaceous plant, 8–28 inches tall; perennial; stems slender, triangular in cross-section, stiff, somewhat enlarged at base; elongate linear leaves (shorter than stem) tapering to a long point, rolled inwardly, arising from near base of plant; inconspicuous flowers covered by dark, glossy brown scales in budlike spikelets borne on slender stalks and forming terminal inflorescence, surrounded and often overtopped by two or three leaflike bracts; fruit dull brown nutlet (achene).

Flowering period. July to October.

Habitat. Irregularly flooded salt marshes and coastal sands.

Range. Long Island to Florida and Texas.

Similar species. Fimbristylis caroliniana also occurs in salt marshes in the same range; its basal sheaths are shorter and thinner, and its bracts are shorter; its scales are short hairy with the midvein forming an elevated keel toward the tip, whereas *F. castanea* has smooth scales without a keel.

× ⅓

Olney Three-square

Scirpus americanus Pers.
(*Scirpus olneyi* Gray)

Sedge Family
Cyperaceae

Description. Medium-height to tall, erect herbaceous plant, up to 7 feet high; perennial from long, hard rhizomes; stems stout and sharply triangular with deeply concave sides; no apparent leaves; inconspicuous flowers borne in five to twelve sessile budlike spikelets covered by brown scales located very near and almost at top of stem (portion above spikelets is ½– 2 inches long); fruit dark gray to black nutlet (achene).

Flowering period. June into September.

Habitat. Irregularly flooded brackish marshes and upper edges of salt marshes; inland saline areas.

Range. New Hampshire and western Nova Scotia to Florida and Mexico; inland in New York, Michigan, and western states, also along the Pacific coast.

Similar species. The stem of *Scirpus pungens* (Common Three-square, formerly *S. americanus*) is stout and triangular but not deeply concaved; its stem is also occasionally twisted. Also, its spikelets are not usually located as close to the top of the stem as they are in *S. americanus*. Other plants with stout, triangular stems have leafy bracts at end of stem, such as *Cyperus strigosus* (Umbrella Sedge), *Scirpus fluviatilis*

(River Bulrush) and *S. robustus* (Salt Marsh Bulrush). The former two species grow in tidal fresh marshes.

× ½

119

Common Three-square or Chair-maker's Rush

Scirpus pungens Vahl
(*Scirpus americanus* Pers.)

Sedge Family
Cyperaceae

Description. Medium-height, erect herbaceous plant, up to 4 feet tall; perennial from hard, elongate rhizomes; stems stout and triangular in cross-section, occasionally twisted; no apparent leaves but actually one to three stemlike erect leaves (up to 16 inches long); inconspicuous flowers borne in several, often three or four, sessile budlike spikelets covered by brown scales located near top of stem (portion above spikelets is 1¼– 5 inches long); fruit gray to black nutlet (achene).

Flowering period. June into September.

Habitat. Brackish and tidal fresh marshes (regularly and irregularly flooded zones) and upper borders of salt marshes where freshwater influence is strong; wet sandy shores, inland marshes, and shallow waters.

Range. Newfoundland, Quebec, and Minnesota, south to Florida and Texas; also in western states to the Pacific coast.

Similar species. The stem of *Scirpus americanus* (Olney Three-square, formerly *S. olneyi*) is triangular and deeply concaved, and its spikelets are located within 2 inches of top of stem. *S. torreyi* (Torrey Bulrush) and *S. smithii* (Bluntscale Bulrush) occur in regularly flooded tidal fresh marshes. The former species has sharply three-angled stems, a soft rhizome, and three-sided nutlets with bristles longer than the nutlet (in *S. pungens* bristles are shorter than nutlet, and in *S. americanus* bristles are of equal length). *S. smithii* has round or bluntly triangular stems. Other plants with triangular, stout stems, e.g., bulrushes (*S. robustus* and *S. fluviatilis*) and umbrella sedges (*Cyperus* spp.), have leafy stems.

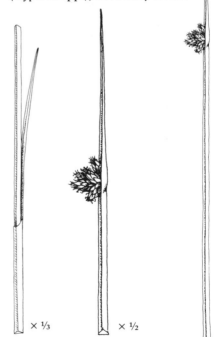

× ⅓   × ½   × 1

Salt Marsh Bulrush

Scirpus robustus Pursh

Sedge Family
Cyperaceae

Description. Medium-height, erect, grasslike herbaceous plant, up to 3 ½ feet tall; perennial with thick rhizome; stems stout and triangular; several elongate, linear, grasslike leaves (½ inch wide) tapering to a long point; inconspicuous flowers borne in three or more spikelets (mostly sessile, few stalked) covered by brown scales (and appearing budlike), inflorescence surrounded by two to four elongate, erect or somewhat erect, leaflike bracts; fruit dark brown to black nutlet (achene).

Flowering period. July to October.

Habitat. Irregularly flooded salt and brackish marshes, occasionally regularly flooded zones.

Range. Nova Scotia to Florida and Texas.

Similar species. Scirpus maritimus (Salt Marsh Bulrush) occurs in salt, brackish, and fresh marshes; it can be distinguishcd from *S. robustus* by the truncate or concave orifice of its leaf sheath, whereas *S. robustus* has a convex orifice. *S. fluviatilis* (River Bulrush) is common in tidal fresh marshes; its leafy bracts are drooping, and most of its spikelets are stalked. Other common brackish bulrushes (*S. americanus* and *S. pungens*) do not have apparent leaves.

leaf sheath

× ½

Scirpus maritimus

× ½

121

Soft-stemmed Bulrush

Scirpus validus Vahl.

Sedge Family

Cyperaceae

Description. Tall, erect herbaceous
plant, up to 10 feet high; perennial
with slender rhizomes; stems soft,
round in cross-section, tapering to
a point, usually grayish green; no
apparent leaves; inconspicuous
flowers borne in an open
inflorescence of many-stalked,
budlike spikelets (⅕–⅘ inch long)
covered by reddish brown scales
located immediately below top of
stem, spikelet clusters mostly
drooping, few erect; fruit brownish
gray nutlet (achene).

Flowering period. June into
September.

Habitat. Brackish and tidal fresh
marshes (regularly and irregularly
flooded zones); inland shallow
waters, shores, and marshes.

Range. Newfoundland to Florida,
west to the Pacific coast.

Similar species. Scirpus acutus
(Hard-stemmed Bulrush) and
S. smithii (Bluntscale Bulrush) also
occur in tidal fresh marshes but are
not common. The former species
has dark green, hard, erect, round
stems, stout rhizomes, and stalked
spikelets forming a nearly terminal
inflorescence. The latter species has
rounded to bluntly triangular stems
and a cluster of spikelets (similar to
S. pungens) about two-thirds up
the stem.

× ½

Baltic Rush

Juncus balticus Willd.

Rush Family
Juncaceae

Description. Medium-height, erect, grasslike herbaceous plant, 1½–3 feet tall; perennial from creeping rhizomes; stems unbranched, round in cross-section, soft, with irregularly vertical fine lines, sheathed at base; no apparent leaves, leaves actually basal sheaths up to 5 inches long with a fine, pointed tip; inconspicuous scaly flowers borne in clusters, either sessile or branched, arising from a single point on the upper half of the stem; fruit capsule.

Flowering period. Late May into September.

Habitat. Sandy salt and brackish marshes (irregularly flooded zone); dunes and inland calcareous shores.

Range. Labrador and Newfoundland to British Columbia, south to Pennsylvania and Missouri; also reported in Harford County, Maryland.

Similar species. Juncus effusus (Soft Rush) is tussock-forming (grows in dense clumps), does not generally occur in brackish marshes, but may be present where freshwater influence is great, and its stem has regular vertical lines, not irregular ones. *J. gerardii* (Black Grass) occurs in salt marshes, but its stem bears conspicuous leaves and is not soft.

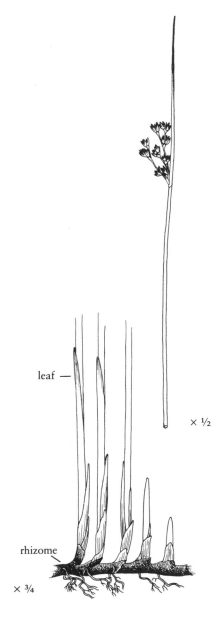

leaf —

× ½

rhizome

× ¾

123

Black Grass

Juncus gerardii Loiseleur

Rush Family

Juncaceae

Description. Low to medium-height, erect, grasslike herbaceous plant, 8–24 inches tall; perennial; one or two elongate linear leaves (up to 8 inches long), round in cross-section, uppermost located near middle of stem; flowers borne on erect or somewhat erect, branched inflorescence (1–3 ¼ inches long); fruit capsule.

Flowering period. June into September.

Habitat. Irregularly flooded salt marshes, usually at upper elevations and sometimes forming cowlicked mats; occasionally, brackish marshes.

Range. Quebec and Newfoundland to Virginia, reported to Florida; also inland in New York, Indiana, and Minnesota; also in Pacific states.

Similar species. Juncus bufonius (Toad Rush) var. *halophilus* occurs in salt marshes from Labrador and Newfoundland to Massachusetts; it is less than 4 inches tall, and its inflorescence makes up more than one-third of the plant's height. The typical form of *J. bufonius* grows to 12 inches high on moist soil throughout the Northeast; its relatively large inflorescence separates it from other coastal wetland rushes. *J. greenei* also occurs in sandy areas (nonwetland) along our coast; its basal leaves are threadlike, nearly round in cross-section, and the only leaf on the stem lies immediately below the inflorescence. Other rushes in salt and brackish marshes have either soft, round, leafless stems (*J. balticus*, Baltic Rush) or round leaves and stems (*J. canadensis* and *J. acuminatus*). *J. roemerianus* (Black Needlerush) is a common brackish marsh plant from southeastern Delaware and Maryland south; its leaves and stems are rigid, round (in cross-section), sharp-pointed, and olive-brown-colored.

× ½

Sweet Gale
Myrica gale L.

Bayberry Family
Myricaceae

Description. Low to medium-height deciduous shrub, up to 5 feet tall, more commonly less than 3 feet tall; twigs brown and usually curving upward; simple, oblong lance-shaped leaves (1¼–2½ inches long) with tapering wedge-shaped bases, entire margins or with a few coarse teeth along outer margin, aromatic when crushed (bayberry scent), alternately arranged; two types of flowers (male and female) borne on separate plants (dioecious) in dense clusters (catkins) at the top of last year's twigs, male catkins elongate cylinder-shaped, female catkins oval-shaped becoming conelike; fruit oval nutlet.

Flowering period. April into June.

Habitat. Upper edges of salt marshes; inland marshes, bogs, swamps, and shallow waters.

Range. Newfoundland and Labrador to Alaska, south to Long Island, New York, and Washington (along coast), and to North Carolina (in mountains in the East).

Similar species. Myrica pensylvanica (Northern Bayberry) has waxy ball-like fruits that often persist through winter and are borne in clusters along twigs below leafy twigs; its leaves are oblong lance-shaped to egg-shaped and generally broader (up to 1½ inches wide) than *M. gale*. Other *Myrica* in coastal wetlands are southern species, and their ranges do not overlap with *M. gale*.

—female

—male

× ¾

125

Bushy Knotweed or Atlantic Coast Knotweed

Polygonum ramosissimum Michx.

Buckwheat Family
Polygonaceae

Description. Low to medium-height, erect herbaceous plant, 1–3½ feet tall; stems jointed, sheathed above joints, and with many ascending branches; simple, entire, linear or narrowly lance-shaped yellow-green, sometimes blue-green, leaves (⅖–2⅗ inches long and less than ⅖ inch wide), tapering at both ends, alternately arranged; small yellow-green flowers, sometimes with pink margins, borne on stalks (pedicels, longer than flowers) from leaf sheaths (ocreae) and greatly dwarfed by leaves; fruit three-sided nutlet (achene).

Flowering period. July through October.

Habitat. Sandy coastal beaches and edges of salt marshes; also, inland dry or moist soils, shores, and roadsides.

Range. Maine and southwestern Quebec to Washington, south along coast to Delaware, inland to Indiana, Texas, and New Mexico.

Similar species. The variety *prolificum* (formerly *Polygonum prolificum*) occurs in salt and brackish marshes from Maine to Virginia; its leaves are blue-green and linear with rounded or somewhat pointed tips and distinct veins that become wrinkled when dry, and its flowers are green with pink or white margins and borne on stalks shorter than the flowers.

× ¾

Marsh Orach or Spearscale
Atriplex patula L.

Goosefoot Family
Chenopodiaceae

Description. Low to medium-height, erect or prostrate, fleshy herbaceous plant, up to 3 ½ feet tall or long; simple, entire, arrowhead-shaped, sometimes narrow or lance-shaped, fleshy leaves (up to 3 inches long) on stalks (petioles), mostly alternately arranged and sometimes oppositely arranged, especially lower leaves; very small green flowers borne in somewhat ball-shaped clusters on open, nearly leafless spikes at upper leaf nodes.

Flowering period. July to November.

Habitat. Irregularly flooded salt and brackish marshes; inland saline or alkaline soils and waste places.

Range. Prince Edward Island and Nova Scotia to British Columbia, south to South Carolina, Missouri, and California.

Similar species. Chenopodium rubrum (Goosefoot) occurs in salt marshes from New Jersey north; it has egg-shaped, coarsely toothed (lobed) leaves that become tinged with red at maturity.

Chenopodium rubrum

× ⅔

× ½

127

Common Glasswort

Salicornia europaea L.

Goosefoot Family
Chenopodiaceae

Description. Low-growing, erect, fleshy herbaceous plant, 4–20 inches tall; annual; stems fleshy, jointed, erect, and much branched or lower branches creeping; leaves reduced to minute scales, blunt or rounded (below spikes), oppositely arranged; inconspicuous green flowers in upper joints of stem forming spikes (less than ⅕ inch wide).

Flowering period. August to November.

Habitat. Irregularly flooded salt marshes, usually in sandy pannes; inland saline soils and marshes.

Range. Quebec and Newfoundland to Florida; inland in New Brunswick, New York, and Michigan; Alaska to California.

Similar species. Salicornia europaea is our most common salt marsh species, but two other glassworts also occur in salt marshes. *S. virginica* (Perennial or Woody Glasswort) has woody, solitary (unbranched) stems. *S. bigelovii* (Bigelow's Glasswort) has sharp-tipped scales below the spikes, thicker spikes (⅕–¼ inch wide), and does not have creeping lower branches.

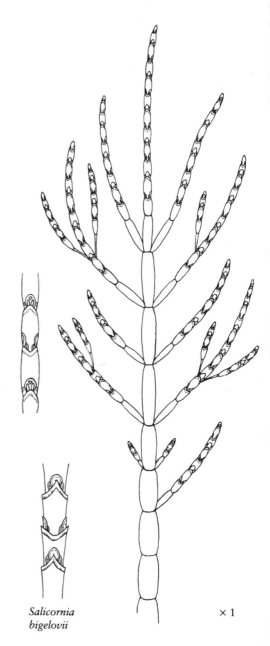

Salicornia bigelovii

× 1

Perennial Glasswort or Woody Glasswort

Salicornia virginica L.

Goosefoot Family
Chenopodiaceae

Description. Low-growing, erect, fleshy herbaceous plant, 4–12 inches tall; perennial; stems somewhat woody-cored, fleshy, and jointed, creeping to form mats, and erect, solitary (unbranched), and flowering; leaves reduced to minute scales, oppositely arranged; inconspicuous green flowers in upper joints of stem.

Flowering period. August into October.

Habitat. Irregularly flooded salt marshes, usually in sandy pannes (salty depressions).

Range. Southern New Hampshire to Florida and Texas; also, Alaska to California.

Similar species. Other *Salicornia* members have nonwoody and branching stems.

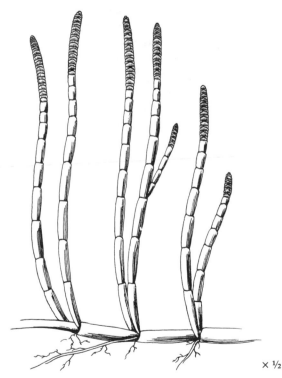

× ½

Saltwort

Salsola kali L.

Goosefoot Family
Chenopodiaceae

Description. Low to medium-height, erect, fleshy and prickly herbaceous plant, 1–3 feet tall; annual; stems smooth or hairy; simple, entire, fleshy, prickly leaves (up to 2 inches long), lower leaves somewhat cylinder-shaped, upper leaves shorter and stiff, with long-spined tip, alternately arranged; small green flowers borne singly or in twos or threes on short spike from axils of upper leaves.

Flowering period. July to October.

Habitat. Beaches and upper edges of irregularly flooded salt marshes, sometimes on top of tidal wrack (vegetation debris—leaves and stems) and also in pannes.

Range. Newfoundland to Florida and Louisiana, rarely inland.

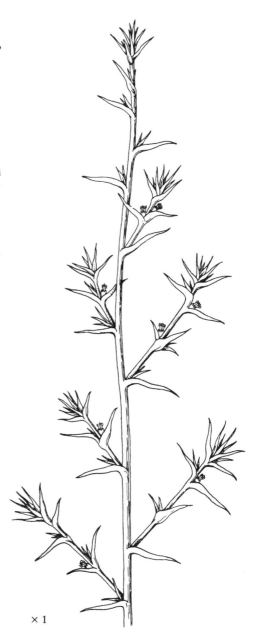

× 1

Sea Blite
Suaeda linearis (Elliott) Moq.

Goosefoot Family
Chenopodiaceae

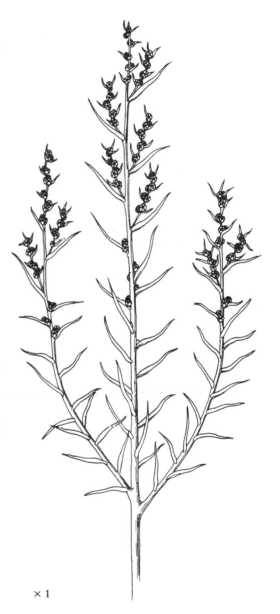

Description. Low to medium-height, erect, fleshy herbaceous plant, 8–32 inches tall; stems usually much branched; simple, entire, dark green, linear fleshy leaves (up to 2 inches long), usually flat on one side and rounded on other, upper leaves reduced in size, alternately arranged; small green flowers borne on terminal spikes, either singly or in clusters of threes in upper leaf axils.

Flowering period. August to October.

Habitat. Irregularly flooded salt marshes, often in sandy pannes.

Range. Maine to Florida and Texas.

Similar species. Other sea blites occur in northeastern salt marshes. *Suaeda maritima* occurs from New Jersey north (rare in Maryland); it is lower-growing, weakly erect or creeping (rarely over 12 inches tall); its leaves are pale green and usually whitened. *S. americana* and *S. richii* occur from Maine north; though both have creeping stems, the former has erect flowering tips and the latter has fleshy leaves rounded on both sides. *Bassia hirsuta* (Hairy Smotherweed) is common from southeastern Maine to New Jersey, rarely to Maryland; its fleshy linear leaves and stems are fine hairy and turn pinkish purple in the fall.

× 1

Water Hemp

Amaranthus cannabinus (L.) Sauer
(*Acnida cannabina* L.)

Amaranth Family
Amaranthaceae

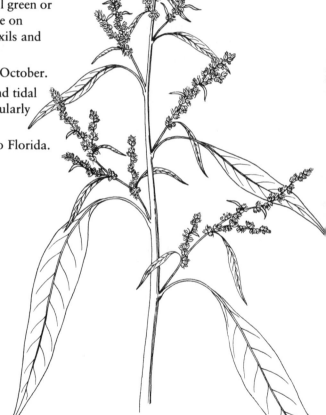

Description. Tall, erect herbaceous plant, 3½–8 feet high; annual; stems smooth; simple, entire leaves (up to 6 inches long), lance-shaped or linear (uppermost), on long stalks (petioles, ⅘–2 inches long), alternately arranged; small green or yellow-green flowers borne on slender spikes from leaf axils and terminally.

Flowering period. July to October.

Habitat. Salt, brackish, and tidal fresh marshes (usually regularly flooded zone).

Range. Southern Maine to Florida.

× ¾

Salt Marsh Sand Spurrey
Spergularia marina (L.) Griseb.

Pink Family
Caryophyllaceae

Description. Low-growing, erect or nearly creeping, fleshy herbaceous plant, up to 14 inches long; annual; stem simple or much branched, smooth or finely hairy, bearing glands; simple, entire, linear fleshy leaves (⅕–1⅗ inches long) with triangular structures (stipules) at leaf bases, oppositely arranged; small pink or white five-petaled flowers (⅙ inch wide) borne on stalks from upper leaf axils.

Flowering period. June through September.

Habitat. Irregularly flooded salt and brackish marshes, usually in sandy pannes; inland alkaline areas.

Range. Quebec to British Columbia, south to Florida and southern California; inland locally to Illinois, Texas, and New Mexico.

Similar species. Spergularia canadensis (Canada Sand Spurrey) occurs in tidal marshes from Newfoundland to New York; its leaves are sharp-tipped, whereas *S. marina* does not have sharp-tipped leaves. *Suaeda linearis* (Sea Blite) and related species have alternately arranged fleshy linear leaves and small green flowers borne on terminal spikes.

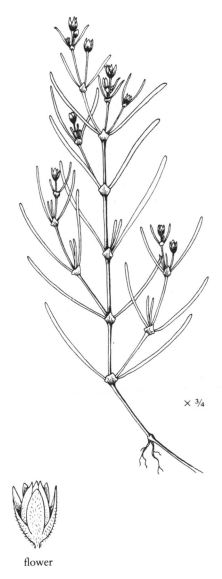

× ¾

flower

× 4

133

Seaside Crowfoot
Ranunculus cymbalaria Pursh

Crowfoot Family
Ranunculaceae

Description. Low-growing, erect, sometimes creeping herbaceous plant, up to 6 inches tall; perennial with spreading rooting branches (stolons); simple, round-toothed, oval heart-shaped to kidney-shaped basal leaves (up to 1 inch long), some leaves not basal; few five-petaled small yellow flowers (up to ⅖ inch wide), with pistils forming dense, conelike head (up to ½ inch long when mature); fruit nutlet (achene).

Flowering period. Summer.

Habitat. Mud flats along regularly flooded brackish and tidal fresh marshes; also inland alkaline muds.

Range. Labrador to Alaska, south to New Jersey, Illinois, and Iowa; also common throughout western states.

Similar species. Ranunculus subrigidus and *R. trichophyllus* (White Water Crowfoots) have submerged, compound leaves divided into threadlike leaflets and have five-petaled white flowers. *R. sceleratus* (Cursed Crowfoot) has small, five-petaled yellow flowers, but its leaves are deeply lobed into three primary parts and then divided again.

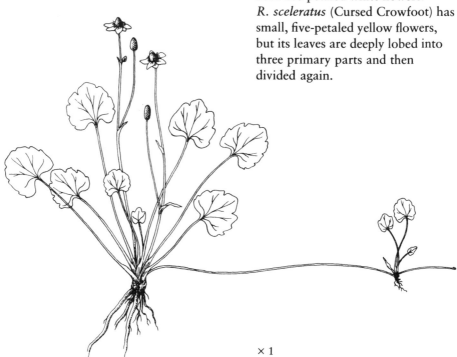

× 1

Sea Rocket
Cakile edentula (Bigel.) Hook.

Mustard Family
Cruciferae (Brassicaceae)

Description. Low-growing, erect, fleshy herbaceous plant, 8–12 inches tall; annual; stems much branched, sometimes creeping; simple, weakly lobed or toothed, sometimes almost entire, fleshy leaves (up to 2 inches long), spoon-shaped, narrowing at the base, alternately arranged; pale purple to white four-petaled flowers (¼ inch wide); fruit two-jointed pod with one or two seeds. *Note:* Fleshy leaves have a mild horseradish taste.

Flowering period. June through September.

Habitat. Upper zone of coastal beaches and upper elevations of salt marshes, usually associated with tidal wrack (vegetation debris—leaves and stems).

Range. Labrador to Florida; inland at the head of Lake Michigan; also reported in Florida.

flower

× 2

× ⅔

Silverweed

Potentilla anserina L.

Rose Family
Rosaceae

Description. Low, creeping, erect herbaceous plant, up to 1 foot tall; perennial; stems creeping; compound basal leaves with seven or more sharply toothed leaflets (up to 1⅗ inches long), silvery hairy beneath, increasing in size toward tip of leaf; yellow five-petaled, sometimes more, flowers (⅗–1 inch wide) borne singly on naked peduncles about as long as leaves.

Flowering period. May through September.

Habitat. Irregularly flooded salt and brackish marshes and wet sandy beaches.

Range. Newfoundland to Alaska, south to New York, northern Indiana and Illinois, Iowa, and in West to New Mexico.

× ½

136

Marsh Mallow

Althaea officinalis L.

Mallow Family
Malvaceae

Description. Medium-height, erect herbaceous plant, 2–4 feet tall; perennial; stems round and smooth hairy; simple, coarsely toothed, soft, velvety hairy leaves (2–4 inches long), egg-shaped, often with three shallow, pointed lobes, alternately arranged; several pink five-petaled regular flowers (1–1½ inches wide) in a peduncled cluster from axils of upper leaves. *Note:* Its thick and mucilaginous roots were the original source of marshmallow.

Flowering period. July through September.

Habitat. Irregularly flooded salt and brackish marshes.

Range. Massachusetts to Virginia and locally inland to Michigan and Arkansas.

Similar species. Kosteletzkya virginica (Seashore Mallow) occurs in salt and brackish marshes from Long Island south; its flowers are somewhat larger (1½–2½ inches wide), its stem is not velvety but fine hairy and rough, and its fruit is a five-celled capsule. *Hibiscus moscheutos* (Rose Mallow) has much larger flowers, often with a red or purple center, a smooth lower stem, and a five-celled fruit capsule.

× ⅓

Rose Mallow

Hibiscus moscheutos L.
(*Hibiscus palustris* L.)

Mallow Family
Malvaceae

Description. Tall, erect herbaceous plant, 3½–7 feet high; perennial; stems round hairy above and smooth below; simple, pointed, toothed leaves, egg-shaped and usually obscurely three-lobed (especially lower leaves) with rounded or heart-shaped bases, smooth above, fine hairy below, alternately arranged; pink or white five-petaled flowers (4–6½ inches wide) with or without purple or red centers; fruit capsule five-celled and rounded at top.

Flowering period. July through September.

Habitat. Irregularly flooded salt, brackish, and tidal fresh marshes; occasionally inland marshes.

Range. Massachusetts to Florida and Alabama; inland from western New York and southern Ontario to northern Illinois and Indiana.

Similar species. Althaea officinalis (Marsh Mallow) and *Kosteletzkya virginica* (Seashore Mallow) are generally shorter (less than 4 feet) and have smaller pink flowers and leaves that are hairy on both sides.

× ½

Seashore Mallow

Kosteletzkya virginica (L.) K. Presl
ex Gray

Mallow Family
Malvaceae

Description. Medium-height, erect herbaceous plant, 2–4 feet tall; perennial; stems round and rough hairy; simple, coarsely toothed, rough hairy leaves (2½–6 inches long), generally triangular egg-shaped, usually with three pointed lobes, alternately arranged; pink five-petaled regular flowers (1½–2½ inches wide) in leaf axils and a terminal spike; fruit capsule five-celled, flattened globe-shaped.

Flowering period. August through September.

Habitat. Irregularly flooded salt and brackish marshes.

Range. Long Island to Florida and Texas.

Similar species. Althaea officinalis (Marsh Mallow) and *Hibiscus moscheutos* (Rose Mallow); see descriptions of these plants for discussion.

× ½

Eastern Lilaeopsis

Lilaeopsis chinensis (L.) Kuntze

Parsley Family
Umbellifereae (Apiaceae)

Description. Very low, erect herbaceous plant, 1–2½ inches tall, growing from creeping rhizome; perennial; stems rhizomatous; simple, flattened, linear basal "leaves" with four to six transverse septa ("leaves" are actually phyllodes—a flattened petiole without a leaf blade); very small white flowers borne in an umbel about as long as "leaves."

Flowering period. June to September.

Habitat. Brackish and tidal fresh marshes (regularly and irregularly flooded zones) and tidal mud flats.

Range. Nova Scotia to Florida and Mississippi.

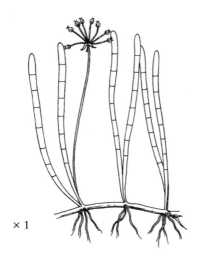

× 1

Mock Bishop-weed
Ptilimnium capillaceum (Michx.) Raf.

Parsley Family
Umbelliferae (Apiaceae)

Description. Low to medium-height, erect herbaceous plant, 8–32 inches tall; annual; compound leaves divided into threadlike leaflets (⅕– 1 inch long), alternately arranged; very small white flowers borne on umbels (⅘–2 inches wide) that overtop leaves.

Flowering period. June through October.

Habitat. Brackish and fresh tidal marshes (regularly and irregularly flooded zones); inland marshes.

Range. Massachusetts to Florida and Texas, chiefly along the coast; inland in southern states to Missouri and Oklahoma; also reported in South Dakota.

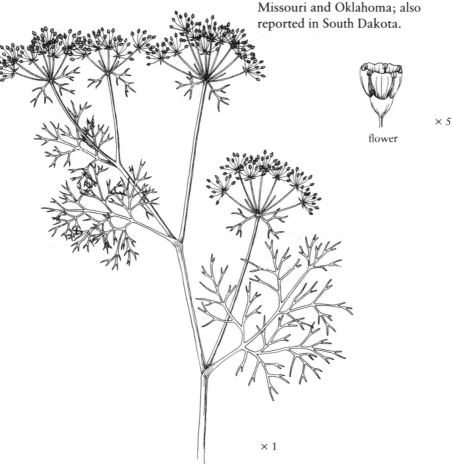

× 5

flower

× 1

Sea Milkwort

Glaux maritima L.

Primrose Family
Primulaceae

Description. Low-growing, erect or creeping, fleshy herbaceous plant, 1¼–14 inches tall; perennial; stems often simple or much branched, erect or prostrate and ascending; simple, entire, sessile, narrowly oblong to linear fleshy leaves (up to ⅘ inch long and ¼ inch wide), round or blunt-tipped, oppositely arranged; small five-lobed pink, white, or red flowers (up to ¼ inch long) borne singly in leaf axils, "petals" joined at base to form a short tube; fruit five-valved capsule bearing few seeds.

Flowering period. June to August.

Habitat. Irregularly flooded salt marshes, often in shallow depressional pannes, and beaches; inland moist or dry alkaline or saline soils.

Range. Eastern Quebec to New Jersey (rarely to Maryland); inland Minnesota to Nevada and New Mexico; on Pacific coast from Alaska to California.

flower

× 3

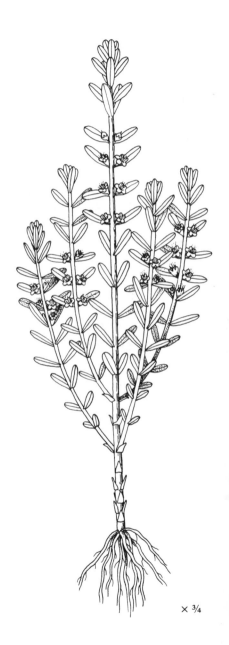

× ¾

Water Pimpernel

Samolus parviflorus Raf.
(*Samolus floribundus* H.B.K.)

Primrose Family
Primulaceae

Description. Low to medium-height, erect herbaceous plant, 4–20 inches tall; perennial; stems branched from upper half and also from base; simple, entire, spoon-shaped or somewhat oval leaves (mostly 1–2 inches long, sometimes to 5 inches), both basal and alternately arranged; small white five-lobed bell-shaped flowers on slender spreading stalks (pedicels, up to ⅘ inch long) with small bract near middle of stalk, borne in terminal inflorescences (racemes, 1¼–6 inches long); fruit round capsule bearing many seeds.

Flowering period. May to September.

Habitat. Brackish and tidal fresh marshes (regularly and irregularly flooded zones) and shallow tidal waters; inland sandy and muddy stream banks and ditches.

Range. New Brunswick to southern Michigan, Missouri, and Kansas, south to Florida and Texas; also in western states.

flower

× 3

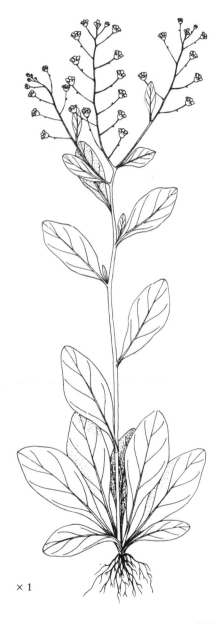

× 1

Sea Lavender or Marsh Rosemary

Limonium nashii Small

Leadwort Family
Plumbaginaceae

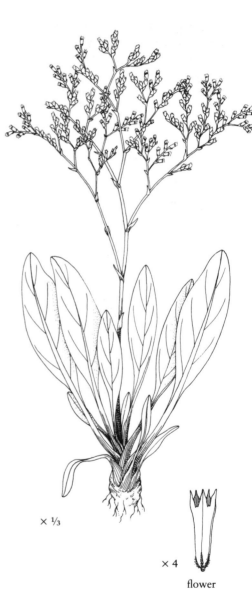

×⅓

Description. Low, erect herbaceous plant with flowering inflorescence, 8–24 inches tall; perennial; simple, entire, basal leaves (2–6 inches long), lance-shaped to spoon-shaped, tapering at base into an often red-tinged petiole usually longer than leaves; numerous minute lavender five-lobed tubular flowers borne on a single, tall inflorescence arising from basal leaves and widely branched above the middle.

Flowering period. July through September.

Habitat. Irregularly flooded salt marshes.

Range. Labrador and Quebec, south to Florida and northeastern Mexico.

Similar species. Limonium carolinianum (Sea Lavender) occurs from southern New York south and rarely to New Hampshire; it has smooth flowers, whereas *L. nashii* has fine hairy flowers (hairy at least at base).

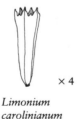

× 4

Limonium carolinianum

× 4

flower

Perennial Marsh Pink or Large Marsh Pink

Sabatia dodecandra (L.) B.S.P.

Gentian Family
Gentianaceae

Description. Low to medium-height, erect herbaceous plant, 1–2½ feet tall; perennial; stems bearing alternate branches above middle; simple, entire, sessile leaves, lance-shaped (⅘–2 inches long), oppositely arranged; pink, sometimes white, regular flowers with eight to twelve petals and yellow center (1½–2½ inches wide).

Flowering period. July into September.

Habitat. Irregularly flooded salt and brackish marshes, rarely tidal fresh marshes.

Range. Connecticut and Long Island, south to Florida and Louisiana.

Similar species. The other common marsh pinks (*Sabatia stellaris* and *S. campanulata*) have only five petals.

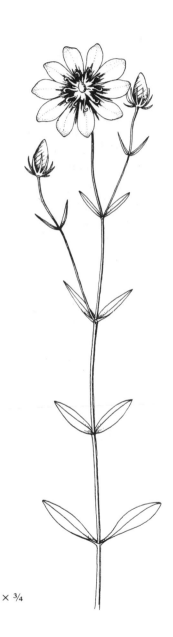

× ¾

145

Annual Marsh Pink

Sabatia stellaris Pursh

Gentian Family
Gentianaceae

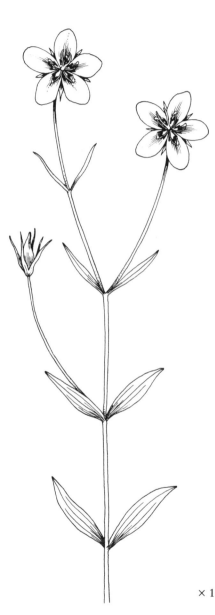

Description. Low to medium-height, erect herbaceous plant, 4–24 inches tall; annual; simple, entire, sessile leaves, linear to egg-shaped, narrow at base (1–1½ inches long), oppositely arranged; pink five-petaled regular flowers with yellow center (¾–1½ inches wide), calyx lobes shorter than petals.

Flowering period. July into October.

Habitat. Irregularly flooded salt and brackish marshes, especially sandy borders of salt marshes.

Range. Southeastern Massachusetts to Florida.

Similar species. Sabatia campanulata (Slender Marsh Pink) occurs in salt and brackish marshes in the same range; its leaves are rounded at base and calyx lobes are as long as petals, whereas *S. stellaris* has leaves narrowing at base and calyx lobes shorter than petals. *S. dodecandra* (Perennial Marsh Pink) has eight to twelve petals.

× 1

Sabatia campanulata

× 1

146

American Germander

Teucrium canadense L.

Mint Family
Labiatae (Laminaceae)

Description. Medium-height, erect herbaceous plant, 1–4 feet tall; perennial; stems square and hairy, rarely branched; sharply or obscurely toothed, simple leaves, oblong to lance-shaped (2–5 inches long) on short stalks (petioles, ¹⁄₁₆–½ inch long), oppositely arranged; small pink-purple or creamy white tubular flowers (⅜–1⅛ inches long) with a broad lower lip (upper lip absent) borne in dense terminal spike.

Flowering period. June through August.

Habitat. Upper edges of salt marshes, irregularly flooded brackish and tidal fresh marshes; inland shores, woods, thickets, and moist or wet soils.

Range. New Brunswick and Nova Scotia, south to Florida and Texas, west to Minnesota and Oklahoma.

Similar species. Mentha arvensis (Wild Mint) is also square-stemmed but has aromatic leaves and stems. It occurs in freshwater tidal marshes but is not found along margins of salt marshes.

× ¾

× 2

flower

147

Seaside Gerardia

Agalinis maritima Raf.
(*Gerardia maritima* Raf.)

Figwort Family
Scrophulariaceae

Description. Low and medium-height, erect herbaceous plant, often 4 inches tall but sometimes to 14 inches; annual; fleshy leaves simple, entire, and linear (up to 1¼ inches long and up to 1/10 inch wide), mostly oppositely arranged but may be alternately arranged on end of branches; small pink to purple five-lobed tubular flowers (½ inch diameter, ⅜–¾ inch long) borne in two to five pairs on stalks (pedicels); calyx lobes blunt with rectangular spaces between. *Note:* Size, number of branches, and number and size of flowers increase north to south.

Flowering period. Mid-July to October.

Habitat. Irregularly flooded salt and brackish marshes, often in pannes.

Range. Nova Scotia to Florida and Texas.

Similar species. Agalinis purpurea (Purple Gerardia, formerly *Gerardia purpurea*) grows to 4 feet tall, its leaves are not fleshy, its flowers are rose-purple in color and larger (¾–1½ inches), the calyx lobes are pointed, and spaces between lobes are V-shaped.

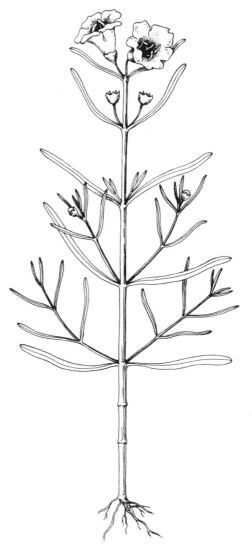

× 1

Mudwort

Limosella subulata E. Ives

Figwort Family
Scrophulariaceae

Description. Low, erect herbaceous plant, up to 2 inches tall; annual; stems prostrate, spreading to form other colonies; simple, entire, linear basal leaves, five to ten in tufts; small white, sometimes tinted with pink, five-lobed tubular flowers borne singly on stalks (peduncles) shorter than leaves; fruit many-seeded round capsule.

Flowering period. Late June through September.

Habitat. Regularly flooded brackish and fresh tidal marshes and intertidal mud and sand flats.

Range. Newfoundland and Quebec, south to Virginia, rarely to North Carolina.

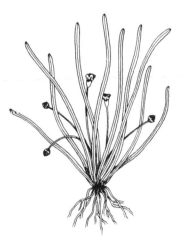

× 1

Seaside Plantain
Plantago maritima L.

Plantain Family
Plantaginaceae

Description. Low-growing, erect, fleshy herbaceous plant, up to 12 inches tall; annual or perennial; simple, mostly entire, fleshy basal leaves (up to 6 inches long), linear to narrowly lance-shaped, tapering to a point; several to many minute greenish or whitish flowers borne on a separate fertile stalk (up to 12 inches long).

Flowering period. June to October.

Habitat. Irregularly flooded salt marshes (common in pannes), beaches, and rocky shores.

Range. Labrador and Newfoundland, south to New Jersey.

Similar species. Fleshy basal leaves of arrow grasses (*Triglochin* spp.) are linear but have conspicuous sheaths.

× ½

Annual Salt Marsh Aster
Aster subulatus Michx.

Composite or Aster Family
Compositae (Asteraceae)

Description. Low to medium-height, erect, somewhat fleshy herbaceous plant, 4–32 inches tall; annual with a short taproot; simple, entire, somewhat fleshy, linear or narrowly lance-shaped leaves (up to 6 inches long), alternately arranged; purplish or blue flowers in heads (less than ½ inch wide) with very short, almost inconspicuous rays, heads usually in an open inflorescence.

Flowering period. August through October.

Habitat. Irregularly flooded salt, brackish, and tidal fresh marshes; inland marshes and thickets and edges of woods.

Range. New Brunswick, southern Maine, and New Hampshire, south to Florida and Louisiana; reported inland near Syracuse, New York, and Detroit, Michigan.

Similar species. Aster tenuifolius (Perennial Salt Marsh Aster) has larger flower heads (½–1 inch wide) and daisylike flowers with longer, more distinct petallike rays; it is a fibrous-rooted perennial with creeping rhizomes.

× ⅔

Perennial Salt Marsh Aster

Aster tenuifolius L.

Composite or Aster Family
Compositae (Asteraceae)

Description. Medium-height, erect, fleshy herbaceous plant, from 6 inches to 2¼ feet tall; perennial with fibrous roots and creeping rhizomes; stems smooth; simple, entire, fleshy, linear, sometimes narrowly lance-shaped leaves (1½–6 inches long), few in number, upper leaves reduced in size, alternately arranged; pale purple or blue or white daisylike flowers in heads (½–1 inch wide) with fifteen to twenty-five petallike rays, several to many heads in an open inflorescence, sometimes solitary.

Flowering period. August through October.

Habitat. Irregularly flooded salt and brackish marshes.

Range. New Hampshire to Florida and Mississippi.

Similar species. Aster subulatus (Annual Salt Marsh Aster) occurs in similar habitats; its flower heads are smaller (less than ½ inch wide), its flowers have very short purplish rays, and it is an annual with a short taproot. *A. simplex* (Lowland White Aster) has white daisylike flowers with twenty to forty petallike rays and is found in tidal fresh marshes, not in salt or brackish marshes.

× ¾

Groundsel Tree or Sea Myrtle

Baccharis halimifolia L.

Composite or Aster Family
Compositae (Asteraceae)

Description. Deciduous shrub, up to 10 feet tall; simple, thick, egg-shaped leaves (up to 2½ inches long), mostly coarsely toothed above middle of leaf, uppermost leaves entire, alternately arranged; white flowers in small heads in stalked (peduncled) clusters forming terminal leafy inflorescences, fertile heads very showy, cottonlike.

Flowering period. August through November.

Habitat. Irregularly flooded salt, brackish, and tidal fresh marshes; open woods and thickets along the coast.

Range. Massachusetts to Florida, Arkansas, and Texas.

Similar species. Sometimes confused with *Iva frutescens* (High-tide Bush) in salt marshes, but *Iva* has opposite and somewhat fleshy leaves.

× 1

Fireweed

Erechtites hieracifolia (L.)
Raf. ex DC.

Composite or Aster Family
Compositae (Asteraceae)

Description. Medium-height to tall, erect herbaceous plant up to 8 feet high; annual; stems somewhat fleshy and marked by fine parallel lines; simple, sharply toothed leaves (up to 8 inches long), lance-shaped or oblong, often fleshy with callous-tipped teeth, alternately arranged; greenish white flowers in heads (½–¾ inch long) with swollen bases and brushlike tips.

Flowering period. July through September.

Habitat. Irregularly flooded salt and brackish marshes; inland marshes, shores, damp thickets, recently burned areas, waste places, and dry woodlands.

Range. Newfoundland to Florida, west to Nebraska and Texas; var. *megalocarpa*—in salt and brackish marshes from Massachusetts to New Jersey.

Similar species. Erechtites hieracifolia grows in two forms; the typical form has normal leaves, while the variety *megalocarpa* has fleshy leaves.

× ¾

Grass-leaved Goldenrod

Euthamia graminifolia (L.) Nutt.
(*Solidago graminifolia* [L.] Salisb.)

Composite or Aster Family
Compositae (Asteraceae)

Description. Medium-height, erect herbaceous plant, 1–4 feet tall; perennial; stems branched at top forming a flattish inflorescence; simple, entire, linear leaves (less than ¼ inch wide), sometimes narrowly lance-shaped, three-nerved (larger leaves with four to five veins), alternately arranged; twenty to thirty-five yellow flowers in heads with fifteen to twenty-five small rays, borne in terminal flat-topped inflorescence.

Flowering period. July through October.

Habitat. Brackish marshes and upper edges of salt marshes; various open moist or dry inland habitats.

Range. Newfoundland and Quebec to Alberta and South Dakota (in mountains to British Columbia), south to Virginia and North Carolina (in mountains to New Mexico).

Similar species. Euthamia galetorum (Slender-leaved Goldenrod, formerly *Solidago tenuifolia*) may also be found in brackish marshes and along salt marsh borders; it has one-nerved leaves and five to seven, rarely nine, yellow flowers with eight to fifteen rays, and it often has clusters of smaller leaves in leaf axils.

× ½

High-tide Bush or Marsh Elder

Iva frutescens L.

Composite or Aster Family
Compositae (Asteraceae)

× ½

Description. Deciduous shrub, 2–12 feet tall, usually less than 6 feet high; stems hairy above and often smooth below; twigs branched with vertical lines; simple, coarse-toothed, somewhat fleshy leaves (2–3 ½ inches long), egg-shaped to narrowly lance-shaped, tapering to a petiole, oppositely arranged except for uppermost reduced leaves; small greenish white flowers in heads borne on erect leafy spikes.

Flowering period. August through October.

Habitat. Irregularly flooded salt marshes, especially on mounds next to ditches and along upper borders.

Range. Nova Scotia and southern New Hampshire, south to Florida and Texas.

Similar species. Sometimes confused with *Baccharis halimifolia* (Sea Myrtle or Groundsel Tree), which also is common in salt marshes, but *Baccharis* has alternately arranged leaves with coarse teeth above the middle.

Annual Salt Marsh Fleabane

Pluchea purpurascens (Swartz) DC.

Composite or Aster Family
Compositae (Asteraceae)

Description. Low to medium-height, erect herbaceous plant, 8–36 inches tall; annual; simple, sharply toothed, hairy aromatic leaves (1 ½–4 ¾ inches long), egg-shaped or lance-shaped with petioles or tapering to the base, alternately arranged; pink or purple flowers in heads (¼ inch long) borne in flat-topped or somewhat rounded inflorescences, bracts hairy.

Flowering period. August through September.

Habitat. Irregularly flooded salt and brackish marshes, occasionally tidal fresh marshes (sometimes regularly flooded zone); interdunal wet swales and inland marshes.

Range. Southern Maine to Florida, west to California; occasionally inland, as in western New York, Michigan, and Kansas.

Similar species. Pluchea foetida (Marsh Fleabane) occurs from southern New Jersey south in freshwater wetlands near the coast; it is perennial with sessile leaves that clasp the stem and creamy white flowers. *P. camphorata* (Camphorweed) occurs from Delaware south; it is quite similar to *P. purpurascens*, but its leaves and stems are usually darker green and smooth and its bracts are smooth or glandular.

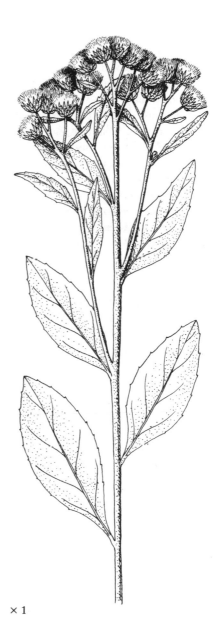

× 1

Seaside Goldenrod
Solidago sempervirens L.

Composite or Aster Family
Compositae (Asteraceae)

Description. Medium-height to tall, erect, fleshy herbaceous plant, usually 3–4 feet high but up to 7 feet; perennial; stems smooth but may be rough hairy in inflorescence; simple, entire, thick, fleshy sessile leaves (4–16 inches long), lance-shaped or oblong, decreasing in size toward top of stem, alternately arranged; numerous yellow flowers in heads with seven to seventeen rays (less than ⅓ inch long) borne on terminal inflorescences.

Flowering period. August through October.

Habitat. Irregularly flooded salt, brackish, and tidal fresh marshes; sand dunes and beaches.

Range. Newfoundland and Quebec to Florida and Texas.

Similar species. Solidago elliottii (Elliott's Goldenrod) also occurs in brackish marshes as well as freshwater swamps from Nova Scotia to Florida; its leaves are toothed and not fleshy. Other salt marsh goldenrods (*Euthamia graminifolia* and *E. galetorum*) have grasslike linear leaves that are not fleshy.

× ½

lower leaf

Plants of Tidal Fresh Marshes

Water Horsetail

Equisetum fluviatile L.

Horsetail Family
Equisetaceae

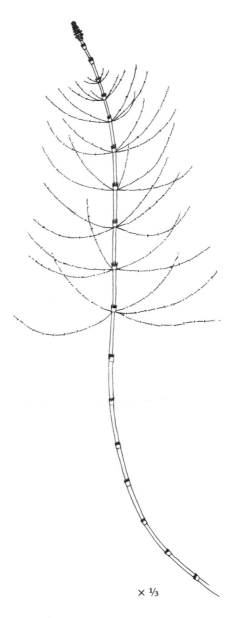

Description. Medium-height, erect herbaceous plant, up to 3½ feet tall; perennial; stems green, seemingly leafless, hollow, jointed, with many vertical ridges and many to no jointed thin branches arranged in whorls below stem sheaths; leaves reduced to minute dark brown, sharp-pointed scales arranged in whorls, fused together forming a collarlike stem sheath; fruits sporangia borne in terminal cone on long stalk.

Fruiting period. May through August.

Habitat. Tidal fresh marshes (regularly and irregularly flooded zones) and shallow waters; inland marshes and shallow waters.

Range. Newfoundland to Ontario, south to Delaware, Pennsylvania, Ohio, and Minnesota; North Dakota to Oregon and Alaska.

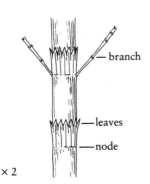

— branch

— leaves

— node

× 2

× ⅓

Riverbank Quillwort

Isoetes riparia Engelm. ex A. Braun
(includes *Isoetes saccharata* Engelm.)

Quillwort Family
Isoetaceae

Description. Low-growing, erect herbaceous plant, 3½–12 inches tall; perennial with stem appearing absent but actually reduced to fleshy bulblike corm; numerous elongate, hollow, erect, linear leaves (up to 12 inches long), sharp-pointed, usually pale green, divided into four air cavities (in cross-section) and separated along leaf length by horizontal cell walls, leaf bases greatly swollen; fruits sporangia borne at base of leaves.

Fruiting period. Summer.

Habitat. Mud flats or gravelly shores along regularly flooded tidal freshwater wetlands; inland shores.

Range. Maine to Virginia.

Similar species. To distinguish from other quillworts requires examination of spores. *Isoetes riparia* is, however, our most common tidal species.

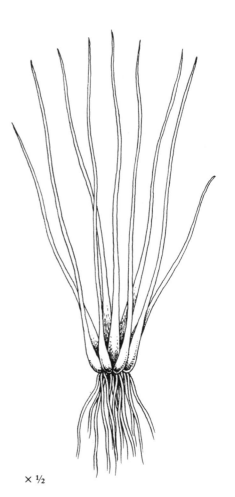

× ½

Sensitive Fern

Onoclea sensibilis L.

Polypody Fern Family
Polypodiaceae

Description. Low to medium-height, erect fern, up to 3½ feet tall; rhizomes brown, usually smooth, and creeping near surface with fibrous rootlets; stalk smooth, thickened at base, yellow with brown; compound light green leaves (fronds up to 14 inches long and 16 inches wide), divided into shallowly lobed leaflets, uppermost connected to one another along stalk, lower leaflets separate; inflorescence separate, arising from rhizome, bearing beadlike fertile leaflets that become dark brown at maturity; fruits sporangia.

Fruiting period. June into October.

Habitat. Tidal fresh marshes; inland marshes, meadows, swamps, and moist woodlands.

Range. Newfoundland to Ontario, Minnesota, and South Dakota, south to Florida and Texas.

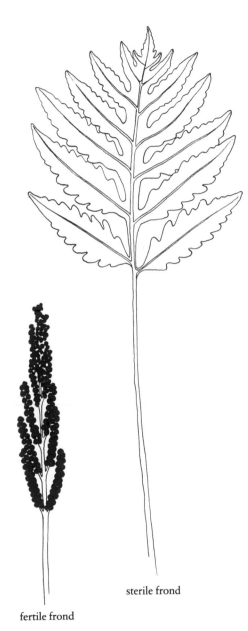

× ⅓

fertile frond

sterile frond

163

Marsh Fern

Thelypteris thelypteroides (Michx.)
J. Holub
(*Thelypteris palustris* Schott)

Polypody Fern Family
Polypodiaceae

Description. Medium-height, erect
fern, up to 28 inches tall; rhizomes
black and branched; stalks about 9
inches long, smooth, slender, and
pale green above and black at base;
compound light green or yellow-
green leaves (fronds up to 16 inches
long and 8½ inches wide), divided
into twelve or more pairs of lance-
shaped leaflets with rounded ends,
two types of leaves (fertile and
sterile), fertile leaves more erect and
on longer stalks than sterile leaves;
fruit dots (sori) borne on undersides
of upper leaflets near midvein.

Fruiting period. June into October.

Habitat. Tidal fresh marshes,
occasionally along upper edges of
salt and brackish marshes; inland
marshes, shrub swamps, and
forested wetlands.

Range. Newfoundland to Ontario
and Manitoba, south to Florida and
Texas.

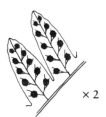

× 2

sori on
underside of
leaflet

× ⅓

Royal Fern

Osmunda regalis L.

Royal or Flowering Fern Family
Osmundaceae

Description. Medium-height to tall, erect fern, 1½–6 feet high, forming tussocks or clumps; rhizomes black and wiry; stalk smooth, straw-colored, and reddish at base; compound leaves (fronds, up to 22 inches wide), twice divided into separate oblong short-stalked leaflets in five to seven pairs per branchlet; fertile leaves with light brown spore-bearing leaflets at the top forming a terminal inflorescence (panicle, up to 12 inches long); fruits sporangia.

Fruiting period. Spring and early summer.

Habitat. Tidal fresh marshes and swamps; inland marshes, swamps, wet meadows, and moist woods.

Range. Newfoundland to Saskatchewan, south to Florida and Texas.

Similar species. Other *Osmunda* have compound leaves (fronds) that are once-divided. *O. claytoniana* (Interrupted Fern) may occur along edges of tidal freshwater wetlands; its spore-bearing leaflets lie between sterile leaflets above and below. *O. cinnamomea* (Cinnamon Fern) is also found in tidal fresh marshes and swamps; its leaves are not divided into separate short-stalked leaflets, and its rhizomes bear separate fertile, cinnamon-colored, woolly leaves (fronds).

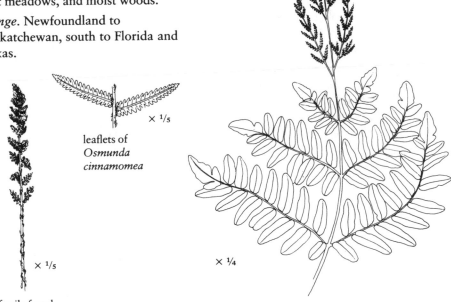

× ⅕

leaflets of
*Osmunda
cinnamomea*

× ⅕

× ¼

fertile frond
of *Osmunda cinnamomea*

165

Great Bur-reed or
Giant Bur-reed

Sparganium eurycarpum Engelm.
ex Gray

Bur-reed Family
Sparganiaceae

Description. Medium-height, erect
herbaceous plant, 1½–5 feet tall;
perennial; simple, entire, stiff linear
leaves (up to 3 feet long) clasping
stem to form sheaths at base,
somewhat triangular in cross-
section, alternately arranged; minute
flowers borne in ball-shaped
heads arranged along a branched
inflorescence; fruit nutlets in ball-
like cluster.

Flowering period. Summer.

Habitat. Tidal fresh marshes;
muddy shores and shallow waters of
rivers, ponds, and lakes, and inland
marshes.

Range. Quebec and Nova Scotia to
British Columbia, south to New
Jersey, Indiana, Kansas, Colorado,
and California.

Similar species. Leaves of
Sparganium americanum (Lesser
Bur-reed) are flat and not stiff;
pistils have one stigma, whereas *S.
eurycarpum* has a pistil with two
stigmas.

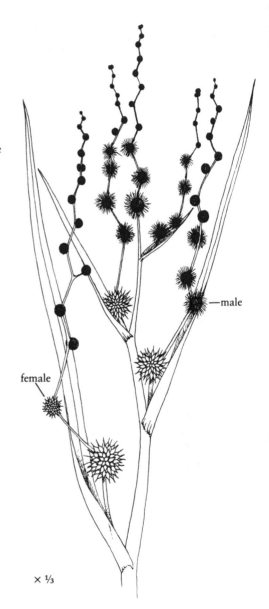

—male

female

× ⅓

166

Northern Water Plantain

Alisma plantago-aquatica L.
(*Alisma triviale* Pursh)

Water Plantain Family
Alismaceae

Description. Low to medium-height, erect herbaceous plant, 4–40 inches tall; perennial from often tuberous rhizome; simple, entire, egg-shaped basal leaves (up to 6 inches long and to 3⅕ inches wide), somewhat heart-shaped at base, usually on long stalks (petioles); numerous small three-petaled white, sometimes pink-tinged flowers (⅕–⅗ inch wide) borne on erect, whorled branched, spreading, terminal inflorescence (panicle, up to 40 inches high); fruits flattened and curved nutlets (achenes) borne in heads (about ⅕ inch wide).

Flowering period. June into September.

Habitat. Regularly flooded tidal fresh marshes and mud flats; inland marshes, shallow waters of ponds, lakes, and streams, and shores.

Range. Nova Scotia and Quebec to British Columbia, south to Maryland, West Virginia, Nebraska, and Arizona.

Similar species. Alisma subcordatum (Southern or Common Water Plantain) usually has smaller and pink or white flowers (less than ⅕ inch wide). *Sagittaria graminea* and *S. falcata* (Grass-leaved Arrowheads) have narrowly lance-shaped or egg-shaped leaves

tapering at both ends and larger three-petaled white, sometimes pink flowers (more than 1 inch wide).

× ¼

Big-leaved Arrowhead
Sagittaria latifolia Willd.

Water Plantain Family
Alismaceae

Description. Medium-height, erect herbaceous plant, from 6 inches to 4 feet tall; perennial; simple, entire, basal leaves (2–16 inches long, 1–10 inches wide), broadly to narrowly arrowhead-shaped; white three-petaled flowers (1–1½ inches wide) arranged in whorls of two to fifteen borne on single elongate stalk (peduncle, up to 4 feet tall); fruits green nutlets (achenes) joined in ball-shaped clusters.

Flowering period. July through September.

Habitat. Tidal fresh marshes; inland marshes and swamps, borders of streams, lakes, and ponds.

Range. Nova Scotia to British Columbia, south to Florida, California, and Mexico.

Similar species. Leaves of *Sagittaria graminea* (Grass-leaved Arrowhead) are flattened and narrowly lance-shaped, not arrowhead-shaped. Leaves of *S. subulata* (Owl-leaf

Arrowhead) and *S. calycina* ssp. *spongiosa* (formerly *Lophotocarpus spongiosus* or *Sagittaria spatulata*) are linear and flattened. All three of these arrowheads occur in regularly flooded brackish and tidal fresh marshes and mud flats.

× ⅓

flower

× ⅙

leaf of
Sagittaria graminea

× ⅓

Sagittaria calycina
ssp. *spongiosa*

× ½

168

Bluejoint

Calamagrostis canadensis (Michx.) Beauv.

Grass Family
Gramineae (Poaceae)

Description. Medium-height to tall grass, 1½–5 feet high; perennial from creeping rhizomes; stems erect, forming clumps; leaves flat (¼–⅘ inch wide), sheaths smooth or mostly so; inflorescence (panicle, 3–8 inches long) loose, open, somewhat drooping, with many branches, lemma with bristle (awn) at middle and surrounded by hairs at base.

Flowering period. June through August.

Habitat. Tidal fresh marshes; inland marshes, shrub swamps, wet meadows, or moist or wet soils.

Range. Greenland to Alaska, south to New Jersey, Delaware, Kentucky, and Missouri, west to Arizona and California; in mountains in the East to North Carolina.

Similar species. Panicum virgatum (Switchgrass) also grows in clumps; its panicle is erect, open and pyramid-shaped, and longer (8–16 inches long). It is typically found along the upland borders of salt marshes and in brackish and tidal fresh marshes.

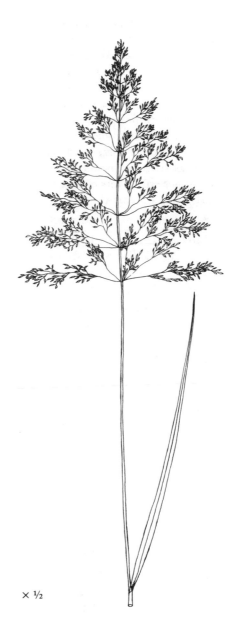

× ½

169

Walter Millet

Echinochloa walteri (Pursh) A. Heller

Grass Family
Gramineae (Poaceae)

Description. Medium-height to tall, erect grass, 3½–6½ feet high; annual; stems round, hollow, and erect; long, tapering leaves (up to 20 inches long and 1 inch wide), leaf sheaths short, coarse hairy; dense terminal inflorescence (panicle, 4–12 inches long) bearing numerous erect spikes with many spikelets covered by long bristles (awns, often 1¼ inches long).

Flowering period. August through October.

Habitat. Tidal fresh marshes; inland fresh and alkaline marshes, swamps, and shallow waters.

Range. Massachusetts to Florida and Texas; inland New York to Wisconsin and Minnesota.

Similar species. Echinochloa crusgalli (Barnyard Grass) is usually less than 3½ feet tall, and its spikelets typically lack bristles (awns).

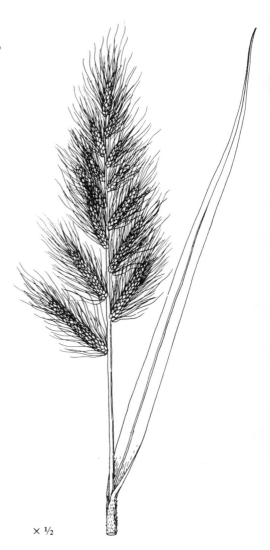

× ½

Virginia Rye Grass
Elymus virginicus L.

Grass Family
Gramineae (Poaceae)

Description. Medium-height to tall, erect grass, 1½–4½ feet high; stems stout and forming clumps; leaf sheaths smooth, leaf blades (⅕–⅘ inch wide) usually rolled inwardly; terminal, unbranched inflorescence (panicle, up to 6 inches long) crowded with flowers (glumes and lemmas) having conspicuously long bristles (awns).

Flowering period. June through August.

Habitat. Tidal fresh marshes and borders of brackish marshes; moist woods, meadows, thickets, shores, and prairies.

Range. Newfoundland to Alberta, south to Florida and Arizona.

Similar species. Echinochloa walteri (Walter Millet) also has spikelets with long bristles (awns), but its panicle is branched.

× ¾

Rice Cutgrass
Leersia oryzoides (L.) Swartz

Grass Family
Gramineae (Poaceae)

Description. Medium-height to tall, erect grass, 2–5 feet high; perennial; stems rough hairy, erect or lying flat on ground at base, then ascending; tapered yellow-green leaves (up to 8 inches long and ½ inch wide), very rough margins with stiff spines, leaf sheaths rough-edged; open terminal inflorescence (panicle, 4–8 inches long) with spreading or ascending slender branches, spikelets (up to ½ inch long) arising from upper half or two-thirds of branches.

Flowering period. June into October.

Habitat. Tidal fresh marshes, occasionally slightly brackish marshes; inland swamps, wet meadows, marshes, ditches, and muddy shores.

Range. Quebec and Nova Scotia to eastern Washington, south to Florida, New Mexico, and California.

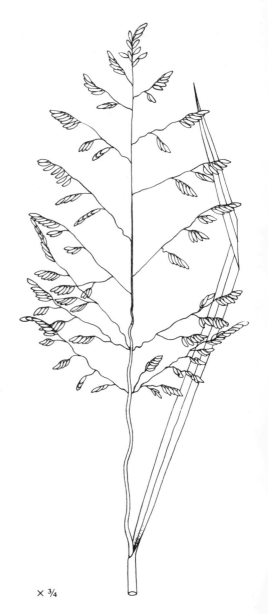

× ¾

Reed Canary Grass
Phalaris arundinacea L.

Grass Family
Gramineae (Poaceae)

Description. Medium-height to tall, erect grass, 2–5 feet high; perennial from creeping rhizomes; stems stout, hollow, round, and erect; tapered leaves (4–12 inches long, ⅖–⅘ inch wide); terminal inflorescence (panicle, 2–7 inches long) branched and compressed early in season but opening after fertilization, spikelets one-flowered.

Flowering period. June through August.

Habitat. Tidal fresh marshes; stream banks, lake shores, marshes, and moist woods.

Range. Newfoundland to Alaska, south to North Carolina, Kansas, and southern California.

× ½

173

Wild Rice
Zizania aquatica L.

Grass Family
Gramineae (Poaceae)

Description. Tall, erect grass up to 10 feet high; annual; stems stout, sometimes lying flat on ground at base, then ascending; large, soft, flat, tapered leaves (up to 48 inches long and 2 inches wide), margins rough; open terminal inflorescence (panicle, 4–24 inches long) divided into two parts, lower part bearing male flowers (first appressed, later ascending) and upper part bearing female flowers (first erect, then open-spreading after fertilization).

Flowering period. June into September.

Habitat. Tidal fresh marshes and slightly brackish marshes (regularly and irregularly flooded zones); stream borders, shallow waters, and inland marshes.

Range. Eastern Quebec and Nova Scotia to Manitoba, south to Florida and Louisiana.

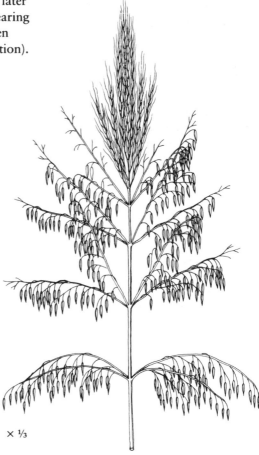

× ⅓

174

Tussock Sedge
Carex stricta Lam.

Sedge Family
Cyperaceae

Description. Medium-height, erect, grasslike herbaceous plant, 1½–3½ feet tall, forming large clumps called tussocks; perennial; stem slender and three-angled; elongate, stiff, linear leaves (up to 2½ feet long, ¼ inch wide) with rough margins, tapering to a tip, channeled and rough above and keeled below, leaf sheath closed at back, leaves arising from base of tussock; inconspicuous flowers borne in two types of spikes—male spikes terminal, one to three in number, and female spikes axillary, two to six in number—covered by overlapping reddish brown or purplish brown scales; fruit nutlet (achene) enclosed by an inflated sac (perigynium).

Flowering period. May through August.

Habitat. Tidal fresh marshes; inland marshes, swamps, and wet swales.

Range. Quebec and Nova Scotia to Minnesota, south to North Carolina, Tennessee, and Iowa.

Similar species. Other common sedges (*Carex* spp.) in tidal fresh marshes are not tussock-forming. They include *C. alata*, *C. albolutescens*, *C. annectens*, *C. comosa*, *C. crinita*, *C. lacustris*, *C. lupulina*, *C. lurida*, *C. normalis*, and *C. rostrata*. Identification of these sedges is difficult and requires examination of achenes and perigynia. Taxonomic references such as Fernald (1970) and Gleason (1952) should be consulted.

× ½

Carex lurida

× ⅟₅₀

tussock

× ¾

Twig Rush

Cladium mariscoides (Muhl.) Torr.

Sedge Family
Cyperaceae

Description. Medium-height, grasslike herbaceous plant, up to 3 feet tall; perennial from rhizomes; stems one or a few, slender and weakly triangular-roundish; elongate, narrow grasslike leaves (up to 8 inches long and ⅟₂₅–⅛ inch wide), grooved near base, rolled inward at tip, weakly rough margins; inconspicuous scale-covered flowers borne in clustered brown spikelets of three to ten, sometimes fifteen to thirty, on upright branches forming inflorescence (up to 12 inches long) from axils of upper leaves and one terminal; fruit nutlet (achene).

Flowering period. August into October.

Habitat. Irregularly flooded brackish and tidal fresh marshes; inland marshes, swamps, shores, and margins of ponds.

Range. Newfoundland and Quebec to Minnesota and Saskatchewan, south to Florida and Alabama.

× ½

Umbrella Sedge or Straw-colored Cyperus

Cyperus strigosus L.

Description. Low to medium-height, erect, grasslike herbaceous plant, 8–40 inches tall; perennial from bulblike rhizome; stems thick, solid, smooth, and triangular in cross-section; elongate linear leaves arranged in three ranks, uppermost leaves (actually bracts) arranged in a cluster at top of stem and immediately below flowering inflorescence; inconspicuous scale-covered flowers borne in cylinder-shaped spikes (½–1½ inches long) forming a terminal inflorescence (umbel), spikes much branched with numerous horizontally radiating or erect flattened yellowish spikelets arranged along slender stalks; fruit three-sided nutlet (achene).

Flowering period. August into October.

Habitat. Tidal fresh marshes; moist fields, swales, inland marshes, swamps, and wet shores.

Range. Quebec and Maine to Minnesota and South Dakota, south to Florida and Texas; also on Pacific coast, Washington to California.

× ⅓

Similar species. Cyperus esculentus (Yellow Nutsedge) and *C. odoratus* (Fragrant Galingale or Flatsedge) also are found in tidal marshes; their spikelets have scales about ⅛ inch long or less. The scales of *C. strigosus* are longer than ⅛ inch and up to ¼ inch. Also, *C. esculentus* is a perennial plant from stolons, whereas *C. odoratus* is an annual plant. Other *Cyperus* spp. of tidal fresh marshes do not have branched spikes but have sessile or stalked unbranched spikes.

Three-way Sedge
Dulichium arundinaceum (L.) Britton

Sedge Family
Cyperaceae

Description. Low to medium-height, erect grasslike plant, 1–3 ½ feet tall, commonly less than 2 feet; perennial; stem hollow, round, and jointed; numerous linear leaves (2–5 inches long and less than ⅓ inch wide), distinctly three-ranked, lower leaves bladeless; inconspicuous flowers covered by scales and borne in several spikes (less than 1 ¼ inches long) with stalks (peduncles, less than 1 inch long); fruit flattened nutlet (achene).

Flowering period. July to October.

Habitat. Irregularly flooded tidal fresh marshes; inland marshes, bogs, swamps, and margins of ponds.

Range. Newfoundland to Minnesota, south to Florida and Texas; also Montana to Washington.

× ¾

Wool Grass
Scirpus cyperinus (L.) Kunth

Sedge Family
Cyperaceae

Description. Medium-height to tall, erect grasslike plant, up to 6½ feet high, commonly 4–5 feet; perennial growing in dense clumps; stem roundish, weakly triangular, especially near base; simple, elongate, rough-margined linear leaves (less than ½ inch wide), drooping at tips, dense cluster of basal leaves present; numerous inconspicuous flowers covered by reddish-brown scales borne on mostly sessile budlike spikelets (usually ⅕ inch long) clustered (three to fifteen) in terminal inflorescence (umbel), somewhat drooping at maturity, and subtended by spreading and drooping leafy bracts; fruit yellow-gray to white nutlet (achene).

Flowering period. August through September.

Habitat. Irregularly flooded tidal fresh marshes; inland marshes, wet meadows, and swamps.

Range. Newfoundland to Minnesota, south to Florida and Louisiana.

Similar species. Scirpus pedicellatus (Wool Grass) has mostly stalked spikelets.

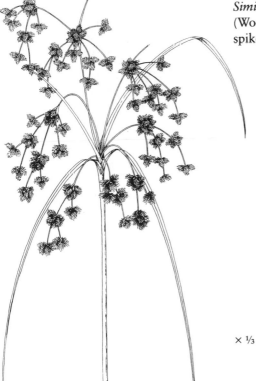

× ⅓

179

River Bulrush

Scirpus fluviatilis (Torr.) Gray

Sedge Family
Cyperaceae

Description. Medium-height to tall, erect, grasslike herbaceous plant, up to 5 feet high; perennial with thick rhizome; stems stout and triangular; elongate, linear grasslike leaves (less than ½ inch wide), leaf sheath orifice convex; inconspicuous flowers borne in spikelets (½–1 inch long) covered by brown scales (and appearing budlike), many spikelets drooping or somewhat erect on long stalks (peduncles), few spikelets sessile, inflorescence terminal with three to five elongate, drooping leaflike bracts below; fruit dull gray-brown three-sided nutlet (achene).

Flowering period. July into September.

Habitat. Tidal fresh marshes; shallow fresh waters of lakes and ponds, inland marshes, and riverbanks.

Range. Quebec to North Dakota, south to Virginia, Indiana, Missouri, and Kansas and from Montana to Washington and California.

Similar species. Scirpus robustus (Salt Marsh Bulrush) occurs in salt and brackish marshes; it has two to four elongate leafy bracts, at least one erect, extending above inflorescence, and its other bracts are somewhat erect, not drooping.

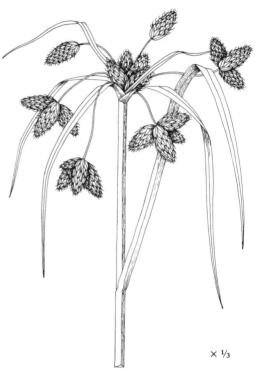

× ⅓

S. maritimus (Salt Marsh Bulrush) occurs in salt, brackish, and tidal fresh marshes; it resembles *S. robustus* but has a truncate or concave orifice of the leaf sheath in further contrast to *S. fluviatilis.*

Sweet Flag
Acorus calamus L.

Arum Family
Araceae

Description. Medium-height, erect herbaceous plant, usually 1−4 feet tall, rarely to 7 feet; perennial; simple, entire, aromatic, linear, sword-shaped basal leaves with midrib slightly off-center; small yellow-brown flowers borne on an erect, 2−4-inch-long fleshy appendage (spadix) developing from a leaflike peduncle (scape).

Flowering period. May to August.

Habitat. Tidal fresh marshes; shallow waters, inland swamps, and wet meadows.

Range. Nova Scotia and Quebec to Montana, Oregon, and Alberta, south to Florida, Texas, and Colorado.

Similar species. Leaves of *Iris* spp. may be confused with *Acorus*, but those leaves are not aromatic.

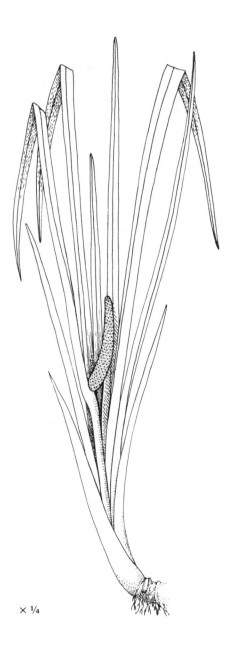

× ¼

Golden Club

Orontium aquaticum L.

Arum Family
Araceae

Description. Medium-height, erect, fleshy herbaceous plant, up to 1½ feet tall; perennial from stout, fleshy rhizome; simple, entire, egg-shaped, fleshy basal leaves (3–10 inches long and about a third as wide) tapering distally to a pointed tip, toward base rolled inwardly where attached to long fleshy stalk (petiole, up to 8 inches long); numerous minute yellow flowers borne at end of separate fertile fleshy stalk (spadix), whitish below and surrounded by a tubular leaf at base.

Flowering period. April through June.

Habitat. Muddy shores of regularly flooded tidal fresh marshes; shallow waters and inland shores.

Range. Massachusetts and central New York, south to Florida and Kentucky.

Similar species. Sagittaria falcata and *S. graminea* (Arrowheads) have somewhat egg-shaped leaves, but their leaves are erect, more narrow, and may be longer (up to 16 inches); their flowers are large, three-petaled, and white and borne in numerous whorls on a tall flowering stalk (scape, to more than 2 feet tall).

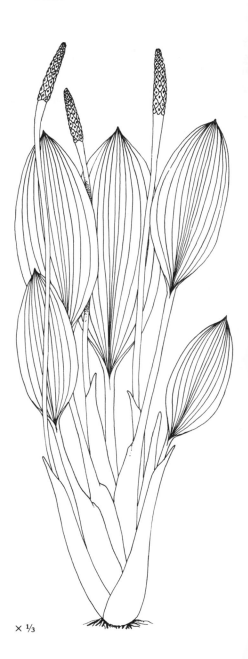

× ⅓

Arrow Arum

Peltandra virginica (L.) Kunth

Arum Family
Araceae

Description. Low to medium-height, erect, fleshy herbaceous plant, up to 2 feet tall; perennial; simple, entire, triangular-shaped fleshy basal leaves (4–12 inches long at flowering and growing larger afterward), ends of basal lobes rounded or pointed, three-nerved, on long petioles; inconspicuous flowers borne on a fleshy spike (spadix) enclosed within a pointed, leaflike green fleshy structure (spathe); fruit greenish, slimy, and pealike berry.

Flowering period. May to July.

Habitat. Tidal fresh marshes and swamps and slightly brackish marshes (regularly and irregularly flooded zones); inland swamps, marshes, and shallow waters of ponds and lakes.

Range. Southern Maine and southwestern Quebec to Michigan, southern Ontario and Missouri, south to Florida and Texas.

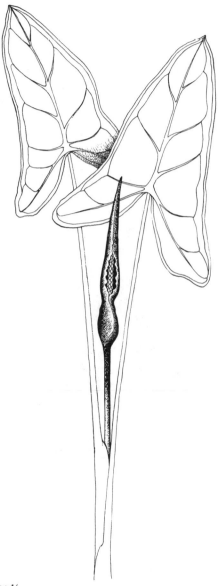

× ¼

Skunk Cabbage

Symplocarpus foetidus (L.) Salisb.

Arum Family

Araceae

Description. Medium-height, erect herbaceous plant, up to 2 feet tall, with enormous foul-smelling (skunklike odor) leaves; perennial from thick rhizome; simple, entire, oval to heart-shaped basal leaves (up to 2 feet long) on short stalks; numerous inconspicuous flowers borne on a fleshy spike (spadix) mostly surrounded by a fleshy, pointed hoodlike structure (spathe), spotted and striped purplish and greenish; fruit berry.

Flowering period. February into May. *Note:* Foul-smelling spadix and spathe present before leaves emerge.

Habitat. Irregularly flooded tidal fresh marshes and swamps; inland marshes, shrub swamps, and forested wetlands.

Range. Quebec and Nova Scotia to southern Manitoba, south to North Carolina and Iowa.

spathe and spadix

× ½

× ⅓

Similar species. *Veratrum viride*
(False Hellebore) also has large
leaves, but they are not foul-
smelling and are deeply ridged, egg-
shaped, sessile, and alternately
arranged in a whorled fashion (three
ranks); its greenish flowers are
borne on a branched terminal
inflorescence (up to 20 inches long),
and its leaves emerge before the
flowers appear.

× ⅓

Veratrum viride

Parker's Pipewort
Eriocaulon parkeri B. Rob.

Pipewort Family
Eriocaulaceae

Description. Low, erect herbaceous plant, 1–4 inches tall; perennial; thin, membranous grasslike basal leaves (1–2½ inches long), linear, tapering to a fine tip; small white flowers in dense button-shaped head at end of four-angled peduncle (scape) that extends above leaves, usually two to four peduncles.

Flowering period. July to October.

Habitat. Tidal freshwater (occasionally slightly brackish) mud flats and shallow waters.

Range. Quebec and Maine to Virginia.

× 1

Mud Plantain

Heteranthera reniformis Ruiz & Pavon

Pickerelweed Family
Pontederiaceae

Description. Low-growing, creeping or weakly erect, fleshy herbaceous plant, up to 6 inches tall; perennial; simple, entire, heart-shaped or kidney-shaped fleshy basal leaves (up to 1¾ inches wide) borne on petioles sheathed at base; two to ten small white or light blue six-petaled star-shaped tubular flowers borne on a spikelike inflorescence sheathed basally by a leafy spathe (from ½– 1¼ inches long); fruit three-valved oval capsule with many seeds.

Flowering period. July through September.

Habitat. Regularly flooded mud flats along edges of tidal freshwater wetlands and shallow tidal fresh waters; inland muddy shores and shallow water.

Range. Connecticut and New York, south to Florida and Texas, west to southern Indiana and Nebraska (in Mississippi valley).

Similar species. Zosterella dubia (Water Star-grass, formerly *Heteranthera dubia*) occurs in similar situations but has linear grasslike leaves and six-petaled yellow tubular flowers borne singly in a leafy spathe.

× 1

× ¼

Zosterella dubia

187

Pickerelweed

Pontederia cordata L.

Pickerelweed Family
Pontederiaceae

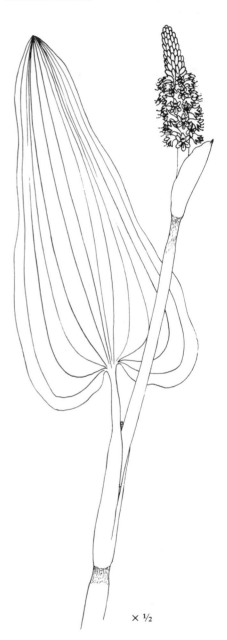

Description. Medium-height, erect, fleshy herbaceous plant, up to 3 ½ feet tall; perennial; simple, entire, heart-shaped, occasionally lance-shaped leaves (up to 7 ¼ inches long) on long petioles, basal and alternately arranged; numerous small violet-blue tubular flowers with three upper lobes (united) and three lower lobes (separated) on terminal spikelike inflorescence (3–4 inches long).

Flowering period. June to November.

Habitat. Tidal fresh marshes, occasionally slightly brackish marshes; inland marshes and shallow waters of ponds and lakes.

Range. Nova Scotia to Ontario and Minnesota, south to northern Florida and Texas.

Similar species. Nuphar luteum (Spatterdock) has heart-shaped leaves with a midrib underneath formed by a continuation of the petiole; it also bears a single yellow flower.

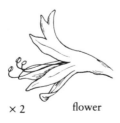

× 2   flower

× ½

Canada Rush

Juncus canadensis J. Gay

Rush Family
Juncaceae

Description. Medium-height, erect grasslike plant, 16–40 inches tall; perennial growing in clumps; elongate linear leaves, round in cross-section with partitions (transverse septa) at regular intervals; five to fifty inconspicuous flowers borne in somewhat rounded heads on compact to open, erect terminal and axillary inflorescences (up to 8 inches long); fruit capsule bearing seeds.

Flowering period. July into October.

Habitat. Slightly brackish and tidal fresh marshes, occasionally borders of salt marshes; inland marshes, swamps, and wet shores.

Range. Quebec to Nova Scotia to Minnesota, south to Georgia and northern Illinois.

Similar species. Juncus acuminatus (Tapertip Rush) has blunt-tipped seeds, whereas *J. canadensis* has seeds with taillike tips.

× ½

189

Soft Rush

Juncus effusus L.

Rush Family
Juncaceae

Description. Medium-height, erect, grasslike herbaceous plant, up to 3 ½ feet tall, forming dense clumps or tussocks; perennial; stems unbranched, round in cross-section, soft, with regular vertical fine lines (ribs), sheathed (usually brown) at base, up to 8 inches long with a bristle tip; no apparent leaves, leaves actually basal sheaths; inconspicuous greenish brown scaly flowers borne in somewhat erect clusters arising from a single point on the upper half of the stem; fruit capsule.

Flowering period. July into September.

Habitat. Tidal fresh marshes; inland marshes, wet meadows, shrub swamps, and wet pastures.

Range. Newfoundland to North Dakota, south to Florida and Texas.

Similar species. Juncus balticus (Baltic Rush) does not form tussocks but grows from creeping rhizomes; its stem has irregular vertical lines (ribs), and it is more common in salt and brackish marshes.

× 1

Wild Yam
Dioscorea villosa L.

Yam Family
Dioscoreaceae

Description. Twining herbaceous vine, up to 16 feet long; perennial from tuberous rhizomes; stems smooth; simple, entire, heart-shaped leaves (2–4 inches long) tapering distally to a slender curved point, with seven to eleven prominent veins, petioled, alternately arranged; small white or greenish yellow flowers borne on short spikes (up to 4 inches long), female flowers borne singly along spike and male flowers borne singly or in clusters of up to four; fruit three-winged capsule.

Flowering period. May through August.

Habitat. Irregularly flooded tidal fresh marshes and swamps; inland shrub swamps, forested wetlands, and roadsides.

Range. Connecticut to Minnesota, south to Florida and Texas.

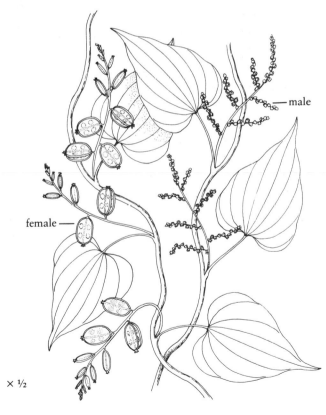

male

female

× ½

191

Blue Flag

Iris versicolor L.

Iris Family
Iridaceae

Description. Medium-height, erect herbaceous plant, 1−3 feet tall; perennial; simple, entire, somewhat fleshy, sword-shaped linear flattened leaves (½−1 inch wide) arising from thick, creeping rhizome in a dense clump, appearing basal but actually alternately arranged and sheathing stem; large blue or violet irislike flowers (2½−4 inches wide) with six "petals" joined in a tube, larger "petals" (sepals) spreading with yellow, green, or white and purple veins, smaller petals erect, borne on a long stalk (8−32 inches tall); fruit bluntly three-angled capsule.

Flowering period. May through July.

Habitat. Tidal fresh and slightly brackish marshes; inland marshes, swamps, wet meadows, and shores.

Range. Newfoundland to Manitoba, south to Virginia and Minnesota.

Similar species. Iris pseudacorus (Yellow Flag) has yellow flowers and valves of dry capsules spread widely at maturity. *Acorus calamus* (Sweet Flag) has sword-shaped linear leaves but is aromatic.

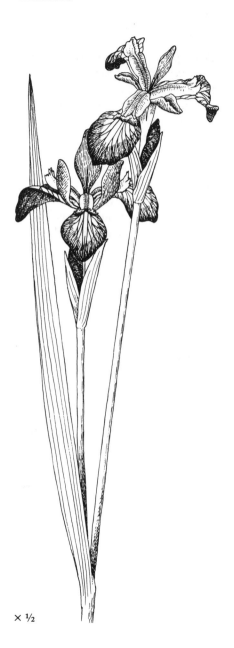

× ½

Lizard's Tail

Saururus cernuus L.

Lizard's Tail Family
Saururaceae

Description. Medium-height to tall, erect herbaceous plant, up to 4 feet high; perennial; stems jointed and slightly branching; simple, entire, somewhat heart-shaped broad leaves (2¼–6 inches long and to 3 inches wide) tapering to a point distally, borne on long petioles and sheathing stem at base, alternately arranged; numerous small white fragrant flowers borne on one or two slender terminal spikes (up to 8 inches long), nodding at tip before all flowers mature; fruit somewhat rounded, wrinkled capsule.

Flowering period. June to September.

Habitat. Tidal fresh marshes and swamps (regularly and irregularly flooded zones); inland swamps, marshes, and shallow waters.

Range. Southern New England, southern Quebec, and Minnesota, south to Florida and Texas.

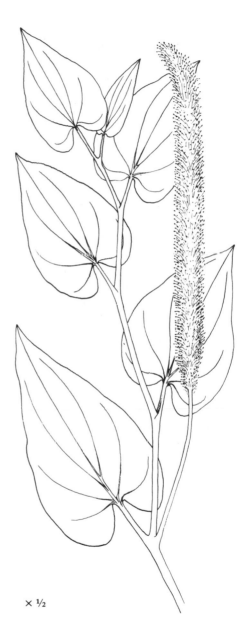

× ½

193

Black Willow
Salix nigra Marshall

Willow Family
Salicaceae

Description. Deciduous shrub or tree, up to 70 feet tall or more; trunk up to 20 inches in diameter with brownish to blackish, deeply grooved bark; yellow-brown to dark brown branchlets, often hairy when young; simple, narrowly lance-shaped, fine-toothed leaves (up to 5 inches long and to ⅘ inch wide) tapering to a long point, with petioles often hairy, alternately arranged, with somewhat heart-shaped, toothed leaflike structures (stipules, up to ½ inch long) at leaf bases; inconspicuous flowers borne on dense spikes (catkins) at end of short, leafy peduncles; fruit somewhat pear-shaped capsule.

Flowering period. April to June.

Habitat. Irregularly flooded tidal fresh marshes and swamps; floodplain forests and meadows.

Range. Southern Canada to central Minnesota, south to Florida and Texas.

Similar species. Salix fragilis (Crack Willow) lacks leafy stipules and has very brittle branchlets that are easily broken at their bases. *S. sericea* (Silky Willow) has leaves with silky hairy undersides, and its flowering spikes appear before the leaves emerge. Leaves of *S. nigra* are green above and light green below (may be hairy when young), and its flowers and leaves appear at the same time.

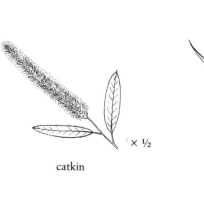

× ½

catkin

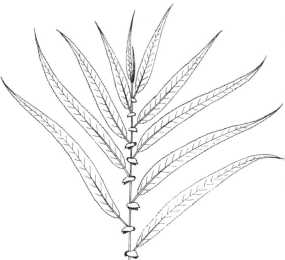

× ¾

194

Wax Myrtle
Myrica cerifera L.

Bayberry Family
Myricaceae

Description. Medium-height to tall evergreen shrub or tree, up to 30 feet high, usually from 10–15 feet; bark smooth and grayish green; twigs waxy with few hairs or hairless; simple, entire or weakly coarse-toothed (above middle), oblong to oblong lance-shaped evergreen leaves (up to 3 ⅘ inches long and ⅘ inch wide), yellow-green, shiny, leathery, aromatic when crushed (bayberry scent), base of leaf wedge-shaped, covered with resin dots (glands) above and below, short-petioled, alternately arranged; two types of flowers (male and female) borne in clusters (catkins) in leaf axils, male catkins linear and female catkins oval; fruit round and waxy ball (drupe).

Flowering period. March into June.

Fruiting period. Summer through winter.

Habitat. Irregularly flooded tidal fresh marshes and swamps, occasionally forming dense thickets, upper edges of salt and brackish marshes, and sandy dune swales; inland swamps (near coast).

Range. Southern New Jersey, south to Florida and Texas.

Similar species. Myrica heterophylla (Evergreen Bayberry) has blackish, soft hairy twigs that are not very waxy; its leaves are not covered by many resin dots on upper surfaces. *M. pensylvanica* (Northern Bayberry) has deciduous leaves that are smooth, dull, and without many resin dots above and often fine hairy below; its young twigs are usually soft hairy and not very waxy. *M. gale* (Sweet Gale) occurs along the coast north of Long Island and does not overlap with *M. cerifera.*

× ½

Smooth Alder

Alnus serrulata (Dryand. in Ait.) Willd.

Birch Family
Betulaceae

Description. Tall deciduous shrub, up to 20 feet or more in height; multiple trunks, dark gray bark marked with small lighter dots (lenticels); twigs dark grayish brown with lenticels and smooth; simple, fine-toothed, egg-shaped leaves (up to 5⅕ inches long and to 3⅕ inches wide), smooth above and usually hairy along veins below, alternately arranged; minute flowers borne in two types of dense spikes (catkins), male catkins elongate (less than 1 inch long), female catkins oval-shaped and appearing conelike after releasing seeds, both spikes persisting through winter; fruit nutlet.

Flowering period. March to May.

Habitat. Irregularly flooded tidal fresh marshes and swamps; inland marshes and swamps.

Range. Maine to New York and Missouri, south to Florida, Texas, and Oklahoma.

Similar species. Alnus rugosa (Speckled Alder) has leaves that are both coarse and fine-toothed (double-toothed), and its bark is marked with large, linear, light-colored marks (lenticels). *A. maritima* (Seaside Alder) is restricted to Delaware and the Eastern Shore of Maryland. Its female catkins are larger (⅘–1⅕ inches long) and borne on stalks, whereas the other two alders have female catkins less than ⅘ inch long, mostly without stalks (attached directly to branch). Also male catkins of *A. maritima* are not present in winter.

× ½

False Nettle or Bog Hemp
Boehmeria cylindrica (L.) Swartz

Nettle Family
Urticaceae

Description. Medium-height, erect herbaceous plant, 1–3 feet tall; perennial; stems unbranched, smooth or rough hairy; simple, coarse-toothed, somewhat broad lance-shaped leaves (1⅕–4⅘ inches long) tapering to a long point distally, petioled, with three distinct veins radiating from leaf base, oppositely arranged; inconspicuous flowers borne in dense elongate spikes borne in leaf axils; fruit shallow-winged oval nutlet (achene).

Flowering period. July through September.

Habitat. Tidal fresh marshes and swamps (regularly and irregularly flooded zones); inland marshes and swamps and moist, usually shaded, soils.

Range. Quebec and Ontario to Minnesota, south to Florida, Texas, and New Mexico.

Similar species. Pilea pumila (Clearweed) looks similar but has translucent stems and glossy leaves.

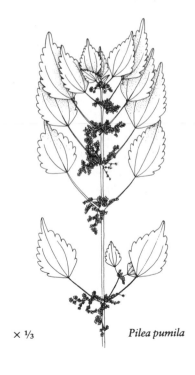

× ⅓   *Pilea pumila*

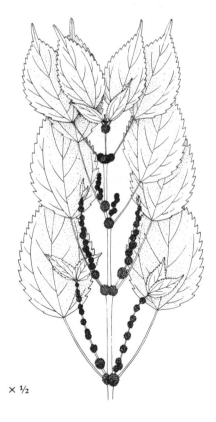

× ½

Halberd-leaved Tearthumb
Polygonum arifolium L.

Buckwheat Family
Polygonaceae

Description. Creeping herbaceous plant, erect when young, up to 4 feet long; stem jointed, weak, several angled, and prickly; simple, entire, hairy leaves (up to 8 inches long), broadly arrowhead-shaped with triangular basal lobes, midrib prickly, alternately arranged; small pink flowers with four lobes in small, close clusters; fruit lens-shaped dark brown to black nutlet (achene). *Note:* Spines on stem point downward.

Flowering period. July through September.

Habitat. Tidal fresh marshes; inland marshes, wet meadows, and swamps.

Range. New Brunswick to Minnesota, south to Florida and Missouri.

Similar species. Polygonum sagittatum (Arrow-leaved Tearthumb) has a four-angled stem, narrowly arrowhead-shaped leaves with rounded bases and mostly pink but sometimes white and green flowers with five lobes.

× ¾

Water Smartweed
Polygonum punctatum Elliott

Buckwheat Family
Polygonaceae

Description. Medium-height, erect herbaceous plant, up to 3 ½ feet tall; annual; stems jointed and sheathed above each joint; simple, entire, smooth leaves (up to 8 inches long), lance-shaped, tapering at both ends, alternately arranged; numerous small green or greenish white flowers arranged in loose, erect spikes, calyx dotted with glands; fruit lens-shaped or three-sided shiny nutlet.

Flowering period. July to October.

Habitat. Tidal fresh marshes (regularly and irregularly flooded zones), occasionally slightly brackish marshes; wet soils, open swamps, and shallow waters.

Range. Quebec to British Columbia, south to Florida and California.

Similar species. Polygonum hydropiper (Common Smartweed) has greenish or red-tipped flowers in loose spikes curving at tips, calyx dotted with glands, and dull lens-shaped or three-sided nutlets. *P. hydropiperoides* (Mild Water Pepper) has mostly pink or purplish flowers, sometimes white or green, in erect, loose cylinder-shaped spikes, and its fruit has a mild peppery taste; also, its calyx is not dotted with glands. *P. pensylvanicum* (Pinkweed) has pinkish or purplish flowers in dense erect spikes.

× ¾

flower
× 4

Arrow-leaved Tearthumb
Polygonum sagittatum L.

Buckwheat Family
Polygonaceae

Description. Creeping herbaceous plant, erect when young, up to 4 feet long; stem jointed, weak, four-angled, and prickly; simple, entire leaves (⅕–4 inches long), lance-shaped with arrowhead-shaped bases or narrowly arrowhead-shaped with somewhat rounded bases, midrib prickly, alternately arranged; small pink, sometimes white or green flowers with five petallike lobes on long stalked heads in leaf axils and terminally; fruit three-angled dark brown or black nutlet (achene). *Note:* Spines on stem point downward.

Flowering period. June through September.

Habitat. Tidal fresh marshes; inland marshes and wet meadows.

Range. Newfoundland and Quebec to Saskatchewan and Nebraska, south to Florida and Texas.

Similar species. Polygonum arifolium (Halberd-leaved Tearthumb) is somewhat similar; see description and illustration.

× ¾

Swamp Dock
Rumex verticillatus L.

Buckwheat Family
Polygonaceae

× ½

Description. Medium-height, erect herbaceous plant, up to 3 ½ feet tall; annual; stems jointed and grooved; simple, entire, narrow lance-shaped flat leaves tapering at base to petiole, alternately arranged; numerous small green flowers, often tinged with red, borne singly on long drooping stalks (pedicels, ⅖ – ⅗ inch long) arranged in whorls along inflorescence (1 – 1 ½ inches long); fruit three-winged.

Flowering period. June to September.

Habitat. Tidal fresh marshes; inland marshes and swamps and edges of streams.

Range. Quebec and Ontario to Wisconsin and Kansas, south to Florida and Texas.

Similar species. Rumex crispus (Sour Dock) has leaves with curled margins.

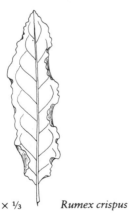

× ⅓   *Rumex crispus*

201

Spatterdock

Nuphar luteum (L.) Sibth. &
J. E. Smith
(*Nuphar advena* Ait.)

Water Lily Family
Nymphaeaceae

Description. Low to medium-height, erect, fleshy herbaceous plant, 8−16 inches tall; simple, entire, heart-shaped, fleshy basal leaves (up to 16 inches long), basal lobes separated by a broadly triangular sinus, borne on rounded stalks (petioles); single yellow flower (1½−2 inches wide) with five or six petals borne on a long fleshy stalk (peduncle).

Flowering period. May to October.

Habitat. Tidal fresh marshes; inland marshes, swamps, and ponds.

Range. Southern Maine to Wisconsin and Nebraska, south to Florida and Texas.

Similar species. Pontederia cordata (Pickerelweed) occurs in the same habitats; it has numerous violet-blue flowers borne on a terminal stalk, and its leaf stalks (petioles) do not form a distinct midrib on the underside of the leaf as in *Nuphar luteum. N. luteum* ssp. *variegatum* (Bullhead Lily) has floating leaves, and its petioles are flattened.

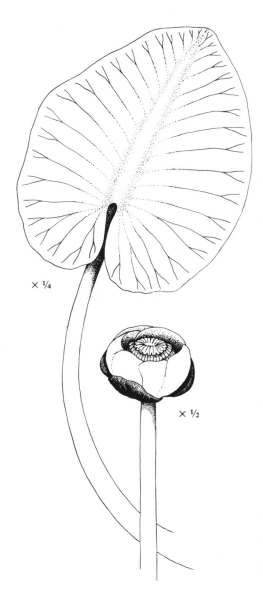

× ¼

× ½

Tall Meadow-rue

Thalictrum pubescens Pursh
(*Thalictrum polygamum* Muhl.)

Crowfoot Family
Ranunculaceae

Description. Tall-growing, erect herbaceous plant, up to 6 feet high, sometimes taller; perennial; stem rigid, smooth or fine hairy; compound sessile leaves, round-toothed leaflets smooth or fine hairy beneath, with main lobes having sharp-pointed tips; numerous white flowers borne in clusters on terminal inflorescence (panicle); fruit six- to eight-winged nutlet (achene).

Flowering period. June through July.

Habitat. Irregularly flooded tidal fresh marshes and swamps; inland marshes, wet meadows, forested wetlands, and stream banks.

Range. Labrador to Quebec and Ontario, south to North Carolina, Tennessee, and Indiana.

× ½

203

Pygmyweed

Crassula aquatica (L.) Schoenl.
(*Tillaea aquatica* L.)

Orpine Family
Crassulaceae

Description. Low-growing, erect, fleshy herbaceous plant, ⅘–4 inches tall; annual; stem branched from base; simple, linear, fleshy sessile leaves (usually ⅓ inch long), oppositely arranged and joined at stem; minute white or greenish white four-petaled flowers borne singly in leaf axils.

Flowering period. July through October.

Habitat. Regularly flooded mud flats along brackish and tidal fresh marshes.

Range. Quebec and Newfoundland to Maryland; also from Louisiana to Texas and along the Pacific coast.

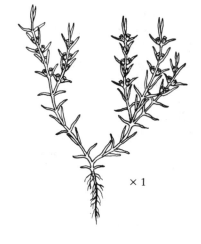

× 1

× 10

flower

Swamp Rose

Rosa palustris Marshall

Rose Family
Rosaceae

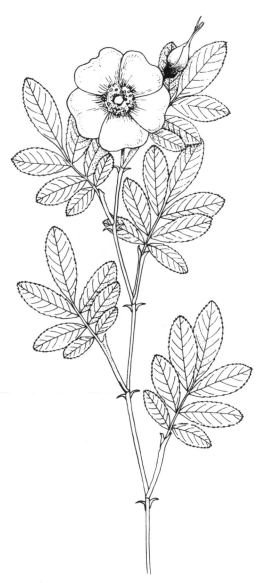

Description. Deciduous thorny shrub, up to 7 feet tall; stems much branched bearing decurved thorns, upper branches smooth except for scattered thorns; compound leaves divided into seven finely toothed, dull green, narrowly egg-shaped leaflets; pink five-petaled flowers (1½–2½ inches wide) borne in small clusters or singly; fruit hairy and red fleshy rose hip enclosing numerous seeds.

Flowering period. June to October.

Fruiting period. Summer into winter.

Habitat. Tidal fresh marshes; inland forested wetlands, shrub swamps, marshes, and stream banks.

Range. Nova Scotia and New Brunswick to Minnesota, south to Florida and Arkansas.

Similar species. Rosa rugosa (Rugosa Rose) grows along salt marshes in thickets and sand dunes from New Jersey north; it has large rose-purple, sometimes white flowers (3–4 inches wide), dark green shiny leaflets, upper branches covered with dense bristles, and larger fruits up to 1 inch wide with leafy tops.

× ½

Sensitive Joint Vetch
Aeschynomene virginica (L.) B.S.P.

Legume Family
Leguminosae

Description. Medium-height to tall, erect herbaceous plant, up to 5 feet high; annual; stems branched and weakly bristle hairy; pinnately compound leaves with odd number of numerous (up to fifty-five) oblong leaflets (⅖–1 inch long) with rounded tips; one to six yellow or reddish flowers with two lips, short tube, and red veins, borne on inflorescences (racemes) in leaf axils; fruit pealike pod with four to ten segments.

Flowering period. August through October.

Habitat. Sandy or muddy tidal shores, tidal fresh marshes (regularly and irregularly flooded zones), and occasionally slightly brackish marshes.

Range. Southern New Jersey to southeastern Virginia, along the coast.

Similar species. Cassia fasciculata (Partridge Pea) also occurs in tidal fresh marshes in Maryland and Virginia but usually is found along the upland border and not on riverbanks. It also has compound leaves; however, its yellow flowers are five-petaled and not red-veined, and its pods are not distinctly segmented.

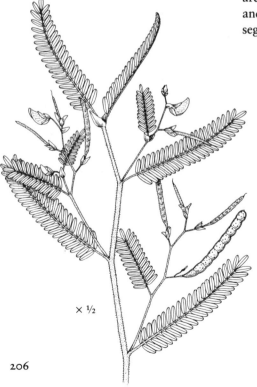

× ½

× ⅔

Cassia fasciculata

False Indigo

Amorpha fruticosa L.

Legume Family
Leguminosae

Description. Deciduous shrub, up to 16 feet tall; smooth dark gray bark; twigs round or finely grooved; alternately arranged compound leaves (up to 16 inches long) divided into eleven to thirty-five narrowly egg-shaped to oblong, short-stalked leaflets (up to 1⅗ inches long and to ⅖ inch wide), somewhat pointed or blunt distally, margins entire, dull green above and usually weakly hairy below; numerous small purplish flowers borne on dense, erect, spikelike inflorescences (racemes, up to 8 inches long); fruits small olive pods marked with red dots.

Flowering period. May into July.

Habitat. Irregularly flooded tidal fresh marshes and swamps; inland moist woods and riverbanks.

Range. New England to Minnesota and Saskatchewan, south to Florida and Texas.

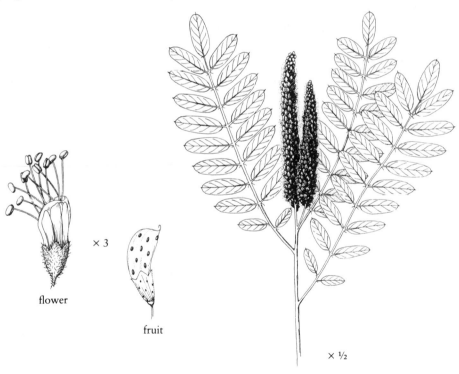

flower   × 3

fruit

× ½

Ground-nut

Apios americana Medic.

Legume Family
Leguminosae

Description. Twining vine; perennial from rhizomes with two or more tubers; compound leaves divided into five to seven (sometimes three) lance-shaped leaflets ($1\frac{3}{5}-2\frac{2}{5}$ inches long) tapering to a fine point distally and somewhat rounded at base; irregular five-petaled (somewhat two-lipped) purplish or brownish fragrant pealike flowers (about ½ inch long) borne singly or in pairs on dense axillary clusters (racemes); fruit linear pod up to 4 inches long.

Flowering period. July into September.

Habitat. Irregularly flooded tidal fresh marshes; moist woods and borders of streams or ponds.

Range. Quebec to Minnesota and South Dakota, south to Florida and Texas.

Similar species. Another twining vine of tidal fresh marshes, *Amphicarpa bracteata* (Hog Peanut), has light purple to white tubular flowers and compound leaves divided into three leaflets.

Amphicarpa bracteata

× ¾

× ½

× ½

Poison Ivy

Toxicodendron radicans (L.) Kuntze
(*Rhus radicans* L.)

Cashew Family
Anacardiaceae

Description. Erect deciduous shrub, trailing vine, or climbing plant up to 10 feet tall when not climbing; twigs brown, older climbing stems densely covered by dark fibers; long-stalked compound leaves (4–14 inches long) divided into three leaflets, end leaflet having a longer stalk than side leaflets, alternately arranged; sap milky; small yellowish flowers with five petals borne on lateral clusters (panicles, up to 4 inches long); fruits small grayish white balls borne in clusters. *Warning:* Do not touch; plant is poisonous and may cause severe skin irritations.

Flowering period. May through July.

Fruiting period. August through November (mostly); some persist through winter.

Habitat. Tidal fresh marshes and along the upper edges of salt marshes; various habitats, mostly dry woods and thickets, but also common in inland wetlands.

Range. Nova Scotia and Quebec to British Columbia, south to Florida and Mexico.

Similar species. Toxicodendron vernix (Poison Sumac, formerly *Rhus vernix*) occurs in tidal fresh marshes and swamps; it is a shrub or small tree (6–23 feet tall) with leaves divided into seven to thirteen leaflets.

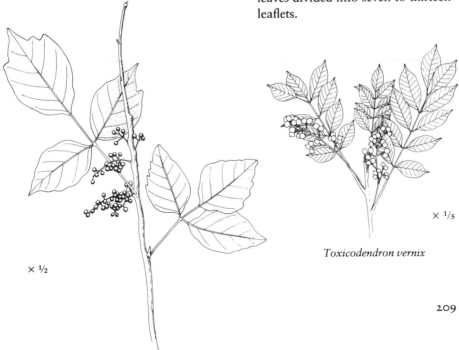

× ½

× ⅕

Toxicodendron vernix

209

Common Winterberry or Black Alder

Ilex verticillata (L.) Gray

Holly Family
Aquifoliaceae

Description. Deciduous shrub, up to 16 feet tall; bark dark gray and smooth; twigs gray and smooth; simple, coarse-toothed, egg-shaped to oblong lance-shaped leaves (2−4 inches long and about 1−2 inches wide) tapering distally to a prominent short point, somewhat wedge-shaped base, petioled, alternately arranged; two types of flowers (male and female) borne on separate plants (dioecious), small white flowers with four to six petals on short stalks borne singly or in clusters (female flowers usually solitary and male flowers in clusters), petals slightly joined at base; fruit bright red, rarely yellow, berrylike (drupe).

Flowering period. May through August.

Fruiting period. Late summer into winter.

Habitat. Irregularly flooded tidal swamps and upper borders of tidal fresh marshes; inland shrub swamps and forested wetlands.

Similar species. Ilex glabra (Inkberry) has smooth, shiny, leathery evergreen leaves with one to three distal teeth on each side margin, small white flowers, and black berrylike fruits.

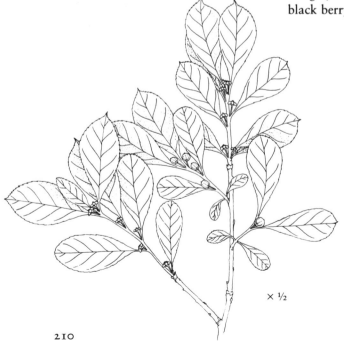

× ½

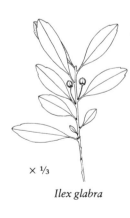

× ⅓

Ilex glabra

Red Maple

Acer rubrum L.

Maple Family

Aceraceae

Description. Deciduous shrub or tree, up to 120 feet tall; smooth gray bark when young, broken and darker when older; young twigs reddish, often partly covered with whitish flaky coating; buds reddish, often clustered near tip of twigs; simple leaves with three to five shallow lobes (2–8 inches long), coarsely toothed, oppositely arranged; small red flowers in short clusters; fruits reddish and winged (samaras). *Note:* The variety *trilobum* has leaves with three lobes.

Flowering period. March through May. *Note:* Red flowers appear before leaf buds open.

Fruiting period. May through July.

Habitat. Tidal fresh marshes and swamps; inland swamps, alluvial soils, and moist uplands.

Range. Newfoundland and Quebec to Manitoba and Minnesota, south to Florida and Texas.

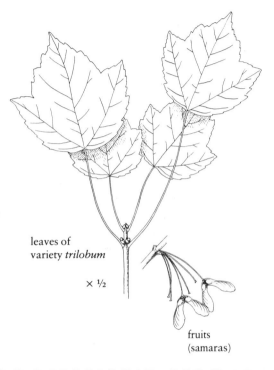

leaves of variety *trilobum*

× ½

fruits (samaras)

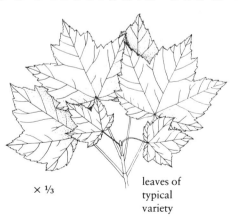

× ⅓

leaves of typical variety

Jewelweed or Spotted Touch-me-not

Impatiens capensis Meerb.

Touch-me-not Family
Balsaminaceae

Description. Medium-height to tall, erect herbaceous plant, 2–5 feet high, rarely 6 feet; annual; stems smooth and succulent; simple, coarsely toothed, soft (almost fleshy) egg-shaped leaves (1⅕–4 inches long) on petioles, alternately arranged; few to several orange or orange-yellow three-petaled tubular flowers (⅘–1⅕ inches long) with reddish brown spots and curved spur at end, borne on long, drooping axillary stalks (pedicels); fruit capsulelike. *Note:* Plants vary in height, leaf size, and color of foliage with differences in exposure and available moisture.

Flowering period. June through September.

Habitat. Tidal fresh marshes, occasionally slightly brackish marshes; inland marshes and swamps, moist woods, stream banks, and springs.

Range. Newfoundland and Quebec west to Saskatchewan, south to Florida, Alabama, Arkansas, and Oklahoma.

× ½

Marsh St. John's-wort

Triadenum virginicum (L.) Raf.
(*Hypericum virginicum* L.)

St. John's-wort Family
Guttiferae (Hypericaceae)

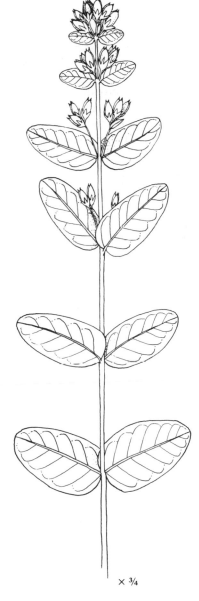

Description. Medium-height, erect
herbaceous plant, 1–2½ feet tall;
stems unbranched, commonly
reddish; simple, entire, oblong to
egg-shaped leaves (1⅕–2⅖ inches
long), usually with round tips and
somewhat heart-shaped bases,
sessile, marked with translucent
dots (glands) beneath, oppositely
arranged; numerous five-petaled
pinkish or purplish flowers (up to
⅘ inch in diameter) borne on
terminal and axillary inflorescences;
fruit tapered capsule.

Flowering period. July through
August.

Habitat. Tidal fresh marshes
(regularly and irregularly flooded
zones), occasionally slightly brackish
marshes; inland marshes, bogs,
swamps, and wet shores.

Range. Nova Scotia to Florida and
Mississippi; inland, New York to
southern Ontario and Illinois.

Similar species. *Hypericum mutilum*
(Dwarf St. John's-wort) also occurs
in tidal fresh marshes; its flowers are
yellow and much smaller (less than
⅕ inch wide).

× 1

flower

× ¾

American Waterwort
Elatine americana (Pursh) Arn.

Waterwort Family
Elatinaceae

Description. Low, creeping or floating herbaceous plant forming mats on mud with branches up to 2 inches long; simple, entire, sessile leaves (less than ⅕ inch long) with rounded tips, broader near distal end and narrowing toward base, oppositely arranged; inconspicuous three-petaled pinkish flowers borne singly in leaf axils; fruit thin-walled capsule.

Flowering period. July to September.

Habitat. Mud flats along tidal fresh and slightly brackish marshes; inland shallow waters and muddy shores.

Range. New Brunswick and Quebec, south to North Carolina, Tennessee, and Texas, west to Washington and California.

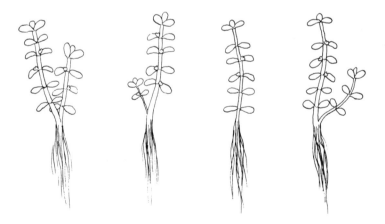

× 1

Water Willow or Swamp Loosestrife

Decodon verticillatus (L.) Elliott

Loosestrife Family
Lythraceae

Description. Medium-height, erect herbaceous plant with somewhat woody base and distinctly arching branches up to 9 feet long and often rooting at tips; stems with four to six angles; simple, entire, narrowly lance-shaped leaves (2–6 inches long), fine-pointed, short-petioled, oppositely arranged or in whorls of threes or fours; numerous five-petaled pink bell-shaped flowers on short stalks borne in clusters in axils of upper leaves; fruit roundish capsule (about ⅕ inch wide) with three to five cells.

Flowering period. July into September.

Habitat. Regularly flooded tidal fresh marshes; borders of rivers, ponds, and lakes, inland marshes, shrub swamps, and forested wetlands.

Range. Central Maine and southern New Hampshire to southern Ontario and Illinois, south to Florida and Louisiana.

× 1

flower

× ½

215

Purple Loosestrife

Lythrum salicaria L.

Loosestrife Family

Lythraceae

Description. Medium-height, erect herbaceous plant, 2–4 feet tall; perennial; stems angled and almost woody; simple, entire, sessile leaves (1 ⅕–4 inches long), lance-shaped, often with heart-shaped bases somewhat clasping stem, oppositely arranged, sometimes in whorls of threes; purple five-to-six-petaled flowers (½–¾ inch wide) borne in dense, leafy, spikelike inflorescences (4–16 inches long).

Flowering period. June through September.

Habitat. Tidal fresh marshes and borders of salt and brackish marshes; inland marshes, wet meadows, and borders of rivers and lakes.

Range. Quebec and New England, west to Michigan and south to Maryland (introduced); native of Eurasia.

Similar species. Lythrum lineare (Salt Marsh Loosestrife) occurs in salt and brackish marshes from Long Island south; it has narrow leaves and small pale purple to white flowers borne in leaf axils, not in dense spikes.

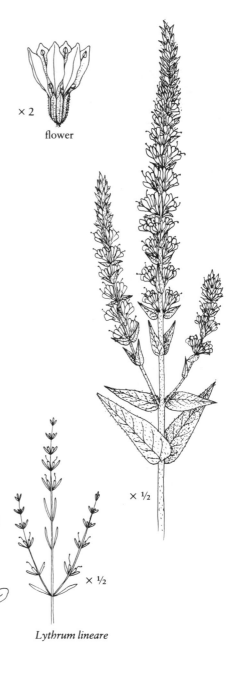

× 2

flower

× ½

× ½

× 2

Lythrum lineare

Seedbox

Ludwigia alternifolia L.

Evening Primrose Family
Onagraceae

Description. Medium-height, erect herbaceous plant, 1½–3½ feet tall; perennial; simple, entire, sessile leaves (2–4 inches long), lance-shaped and pointed, alternately arranged; small four-petaled yellow flowers (½–¾ inch wide) borne singly in leaf axils; fruit capsule square at top, rounded at base, with a terminal pore.

Flowering period. June through August.

Habitat. Tidal fresh marshes; inland marshes, swamps, and wet soils.

Range. Massachusetts and southern Ontario to Iowa and Kansas, south to Florida and Texas.

× 1

× 1½

flower

Water Parsnip

Sium suave Walter

Parsley Family
Umbelliferae (Apiaceae)

Description. Tall, erect herbaceous plant, up to 7 feet high; perennial; stems grooved or strongly angled and smooth; compound leaves with seven to seventeen leaflets (1–4 inches long), linear or lance-shaped, strongly toothed, upper leaflets often simple, alternately arranged; very small white flowers borne in umbels (1⅕–4⅘ inches wide); fruit somewhat elongate oval capsule with prominent ribs.

Flowering period. July through September.

Habitat. Slightly brackish marshes and tidal fresh marshes; inland marshes, swamps, and muddy shores.

Range. Newfoundland to British Columbia, south to Florida, Louisiana, and California.

Similar species. Cicuta maculata (Water Hemlock) occurs in freshwater marshes; its leaves may be once, twice, or thrice divided, some leaflets are three-lobed, and its stem is not strongly angled and may be purple mottled. *Caution:* Its fleshy roots are extremely poisonous.

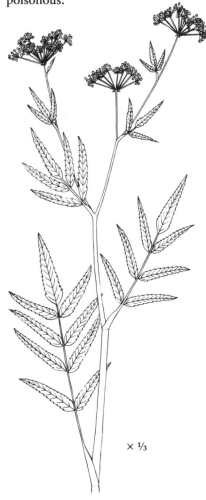

× ¼   *Cicuta maculata*   × ⅓

Silky Dogwood
Cornus amomum Mill.

Dogwood Family
Cornaceae

Description. Much branched deciduous shrub, 3½–10 feet tall; bark gray and reddish purple striped; twigs dull reddish purple and silky hairy, with light brown pith; simple, entire, egg-shaped leaves (2–5 inches long) with rounded bases and three to five pairs of veins, oppositely arranged; numerous small white four-petaled flowers borne on flat-topped or somewhat rounded inflorescence at tips of branches; fruit bluish or bluish white, berrylike (drupe) with stone seed.

Flowering period. May through July.

Fruiting period. August through October.

Habitat. Tidal fresh marshes; moist or wet thickets and woods, borders of inland marshes and streams.

Range. Southern Maine to southern Indiana and Illinois, south to Georgia and Alabama.

Similar species. Cornus stolonifera (Red Osier) has glossy red bark and branches with white pith; its fruits are white, rarely with bluish tinge.

× 2

flower

× ⅔

Sweet Pepperbush

Clethra alnifolia L.

White Alder Family
Clethraceae

Description. Deciduous shrub, up to 10 feet tall; bark grayish brown and flaky; simple, coarse-toothed, somewhat egg-shaped to oblong leaves (2–4 inches long) tapering to a fine point distally and wedge-shaped near base, toothed along upper leaf margin, lower margin mostly entire, petioled, alternately arranged; numerous small five-petaled fragrant white flowers borne on terminal spikelike inflorescences (racemes); fruit hairy three-valved capsule (persists through winter).

Flowering period. July through September.

Habitat. Irregularly flooded tidal swamps and upper edges or higher elevations within tidal fresh marshes; inland forested wetlands, shrub swamps, and sandy woods.

Range. Southern Maine to Florida and Texas.

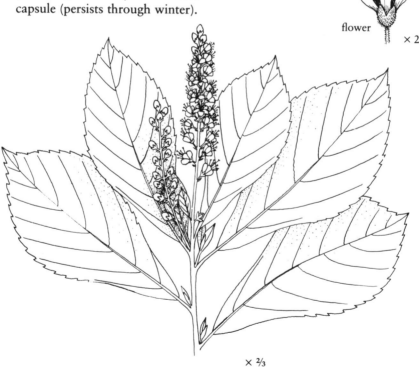

flower   × 2

× ⅔

Green Ash

Fraxinus pennsylvanica Marshall
var. *subintegerrima* (Vahl) Fernald

Olive Family
Oleaceae

Description. Deciduous shrub or tree, up to 80 feet tall; brown, shallowly grooved bark on trees; twigs gray and smooth; oppositely arranged, compound leaves divided into five to nine, usually seven, shallow-toothed lance-shaped leaflets (up to 6 inches long) tapering to a blunt or fine point distally; flowers inconspicuous; fruits winged (samaras).

Flowering period. April to June. *Note:* Flowers appear as leaves emerge.

Fruiting period. Summer to fall.

Habitat. Tidal swamps and higher areas within and borders of irregularly flooded tidal fresh marshes; inland swamps.

Range. Maine to Ontario and Saskatchewan, south to Florida and Texas.

Similar species. Fraxinus pennsylvanica (Red Ash) in its typical form has velvety hairy twigs, light brownish hairy petioles and lower leaf surfaces.

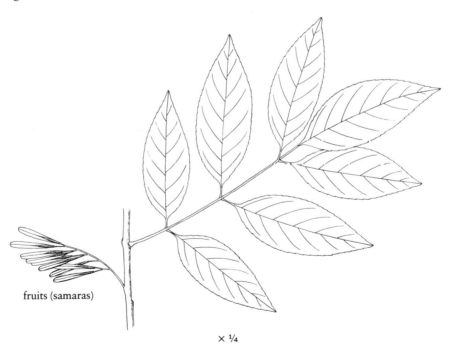

fruits (samaras)

× ¼

221

Swamp Milkweed

Asclepias incarnata L.

Milkweed Family
Asclepiadaceae

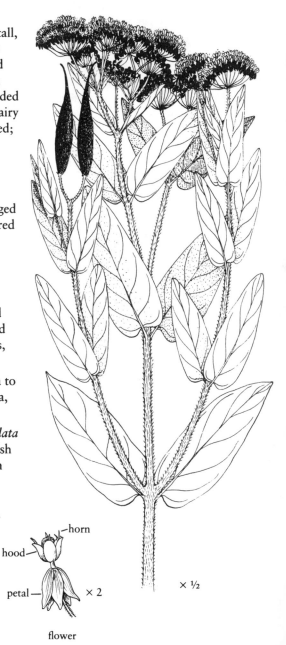

Description. Medium-height to tall, erect herbaceous plant, 2–6 feet high; perennial; stems round and smooth or hairy, with milky sap; simple lance-shaped leaves, rounded or tapering at base, smooth or hairy on both sides, oppositely arranged; numerous pink to purplish red regular flowers composed of five erect hoods (⅕ inch long) with somewhat longer horns and five downward-pointing lobes, arranged in several umbels; fruit pod tapered at both ends and standing erect from branches.

Flowering period. June through August.

Habitat. Tidal fresh marshes and edges of brackish marshes; inland shrub swamps, forested wetlands, marshes, shores, and ditches.

Range. Nova Scotia to Manitoba to Utah, south to Florida, Louisiana, and New Mexico.

Similar species. Asclepias lanceolata (Red Milkweed) occurs in brackish and fresh marshes from southern New Jersey south; its leaves are linear or narrowly lance-shaped, and its flowers are orange or red.

horn

hood

petal

× 2

× ½

flower

Hedge Bindweed

Calystegia sepium (L.) R. Br.
(*Convolvulus sepium* L.)

Morning Glory Family
Convolvulaceae

Description. Twining and sometimes trailing herbaceous vine, up to 10 feet long; simple, entire, triangular-shaped leaves (2−4 inches long), often with somewhat squarish basal lobes, on long petioles, alternately arranged; large white or pink funnel-shaped tubular flowers (1½−3 inches long) borne usually singly on long stalks (peduncles, 2−6 inches long).

Flowering period. Mid-May into September.

Habitat. Tidal fresh marshes, occasionally brackish marshes and beaches; inland moist thickets, shores, roadsides, and waste places.

Range. Quebec and Newfoundland to British Columbia, south to Florida, Missouri, and Oregon.

× ½

Common Dodder

Cuscuta gronovii Willd.
ex J. A. Schultes

Morning Glory Family
Convolvulaceae

Description. Slender, parasitic, twining herbaceous nonleafy vines; stems smooth and orange or yellow-orange; leaves apparently lacking but actually reduced to minute scales; small white or yellowish bell-shaped flowers (1/10 – 1/5 inch long) in sessile clusters; fruit ball-shaped capsule.

Flowering period. July to October.

Habitat. Tidal fresh marshes, occasionally in salt and brackish marshes, especially parasitizing high-tide bush (*Iva frutescens*); inland marshes and other low areas.

Range. Nova Scotia to Manitoba, south to Florida, Texas, and Arizona.

Similar species. To distinguish *Cuscuta gronovii* from other dodders, one must closely examine flowers and follow a technical taxonomic reference. *C. gronovii* is our most common dodder.

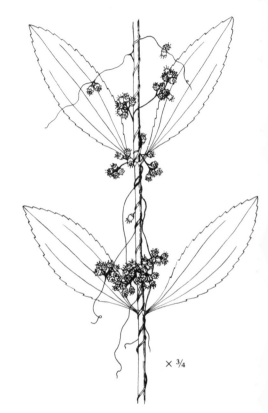

× 3/4

Host plant is *Iva frutescens*

224

Water Horehound or Bugleweed

Lycopus virginicus L.

Mint Family
Labiatae (Lamiaceae)

Description. Medium-height, erect herbaceous plant, up to 3 feet tall; perennial, usually from tuberous root; stems four-angled, usually fine hairy, often producing long runners from base; simple, coarse-toothed, lance-shaped leaves (2–5 inches long) tapered at both ends, with coarse marginal teeth beginning just below middle of leaf, leaves generally dark green, sometimes purple-tinged, oppositely arranged; numerous small four-petaled white tubular flowers borne in dense ball-like clusters at leaf bases; fruit nutlet.

Flowering period. July into October.

Habitat. Irregularly flooded tidal fresh marshes; inland marshes, wet meadows, and forested wetlands.

Range. Quebec and Nova Scotia to Minnesota, south to Georgia, Alabama, and Oklahoma.

Similar species. Lycopus americanus (Water Horehound) also occurs in tidal fresh marshes; its lower leaves are deeply lobed, and it lacks a tuberous root. *Mentha arvensis* (Wild Mint) looks similar but has a strong minty odor.

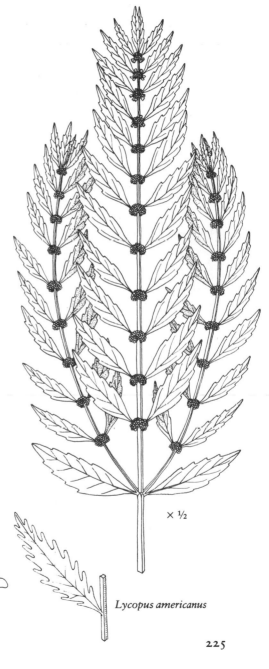

× ½

flower

× 5

Lycopus americanus

225

Wild Mint

Mentha arvensis L.

Mint Family
Labiatae (Lamiaceae)

Description. Low to medium-height, erect or ascending herbaceous plant, 1–1 ½ feet tall or sometimes more; perennial; stems four-angled, hairy on angles and smooth or hairy on sides; simple, coarse-toothed, lance-shaped to oblong leaves (1 ⅖ – 2 ⅕ inches long) on petioles, aromatic (leaves have strong minty odor when crushed), oppositely arranged; numerous minute light blue, lavender, or white two-lipped tubular flowers borne in dense ball-like clusters in leaf axils along the upper half or third of the stem.

Flowering period. July through September.

Habitat. Irregularly flooded tidal fresh marshes; inland marshes and moist soils.

Range. Labrador to Alaska, south to Virginia, Kentucky, and Missouri, west to California.

Similar species. Lycopus spp. (Water Horehounds) have four-angled (square) stems and numerous small white tubular flowers borne in dense ball-like clusters in leaf axils, but they do not have a minty odor.

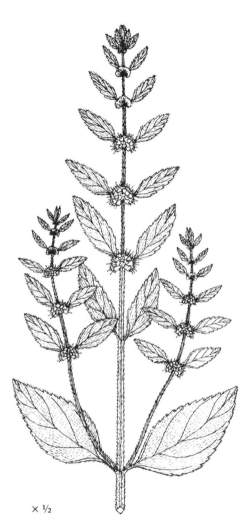

× ½

Mad-dog Skullcap

Scutellaria lateriflora L.

Mint Family
Labiatae (Lamiaceae)

Description. Medium-height, erect herbaceous plant, 1–2½ feet tall; perennial from slender rhizomes; stems four-angled, slender, usually branched, smooth or fine hairy on angles; simple, coarse-toothed, broadly lance-shaped leaves (up to 3 inches long) tapered to a point distally, rounded at base, petioled, oppositely arranged; numerous small blue, sometimes pink or white, two-lipped tubular flowers borne on one side of inflorescences (racemes) in leaf axils and usually one terminal, flowers usually subtended by small lance-shaped leaves.

Flowering period. June through September.

Habitat. Irregularly flooded tidal fresh marshes; inland marshes, wet meadows, and swamps.

Range. Quebec to British Columbia, south to Florida, Louisiana, and Arizona.

Similar species. Scutellaria galericulata (Common Skullcap, formerly *S. epilobiifolia*) has blue flowers borne singly from leaf axils and short-petioled or stalkless leaves.

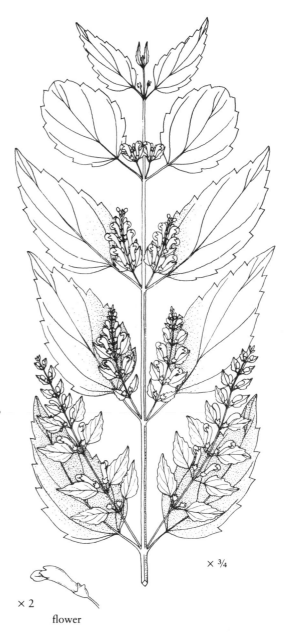

× ¾

× ⅔

Scutellaria galericulata

× 2

flower

Overlooked Hedge Hyssop or Clammy Hedge Hyssop
Gratiola neglecta Torr.

Figwort Family
Scrophulariaceae

Description. Low-growing, erect herbaceous plant, 4–12 inches tall; annual; stem unbranched or widely branched, upper part hairy and sometimes sticky; simple, coarse shallow-toothed, lance-shaped to oblong lance-shaped leaves (⅘–2½ inches long) tapering at both ends, most leaves sessile but young ones stalked, oppositely arranged; small yellow tubular flowers with five white lobes borne on somewhat drooping slender stalks up to 1 inch long; fruit roundish capsule.

Flowering period. May to October.

Habitat. Regularly flooded mud flats along edges of tidal fresh marshes; inland shores and shallow waters.

Range. Quebec to southern British Columbia, south to Georgia, Texas, and Arizona.

Similar species. *Gratiola virginiana* (Virginia Hedge Hyssop) has white tubular flowers internally marked with purple lines borne on short erect stalks. *G. aurea* (Golden-pert) has bright yellow tubular flowers, somewhat four-angled stems, and somewhat clasping, simple, usually entire leaves. *Lindernia dubia* (False Pimpernel) has light purplish or whitish tubular flowers with five lobes forming two lips, upper lip two-lobed with shallow notch, lower lip three-lobed.

× ⅓

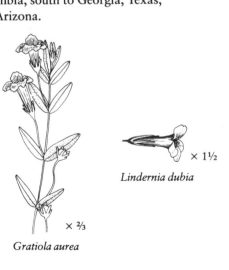

× 1½

Lindernia dubia

× ⅔

Gratiola aurea

Square-stemmed Monkeyflower

Mimulus ringens L.

Figwort Family
Scrophulariaceae

Description. Medium-height, erect herbaceous plant, up to 3 ½ feet tall; perennial; stems four-angled or very narrowly winged and smooth; simple, obscurely toothed, sessile, lance-shaped or narrowly oblong leaves, larger leaves 2–4 inches long, oppositely arranged; small blue to purple tubular flowers (1–1 ½ inches long) with two lips, upper lip two-lobed and lower lip three-lobed, borne in leaf axils on long stalks (peduncles, ⅘–1⅕ inches long); fruit many-seeded capsule.

Flowering period. Late June through September.

Habitat. Tidal fresh marshes; inland marshes, wet meadows, and shores.

Range. Quebec and Nova Scotia to Saskatchewan, south to Georgia, Alabama, Louisiana, and Oklahoma.

Similar species. Mimulus alatus (Sharp-winged Monkeyflower) has petioled (stalked) leaves, and its flower stalks (peduncles) are short (less than ⅗ inch long).

× ½

Buttonbush

Cephalanthus occidentalis L.

Madder Family
Rubiaceae

Description. Deciduous shrub, 3½–
10 feet tall; young bark smooth and
grayish, older bark grayish brown
and flaky; pith light brown; twigs
grayish brown to purplish, round,
hairy or smooth, and marked with
light elongate dots (lenticels);
simple, entire, egg-shaped leaves
(3–6 inches long) tapering to a
short point, oppositely arranged
but sometimes in whorls of threes
and fours, leaf stalks often red;
small white tubular flowers in dense
ball-shaped heads; fruit nutlet-
bearing ball.

Flowering period. May through
August.

Fruiting period. September through
December.

Habitat. Tidal fresh marshes; inland
marshes, shrub swamps, forested
wetlands and borders of streams,
lakes, and ponds.

Range. New Brunswick and Quebec
to Minnesota, south to Florida,
Mexico, and California.

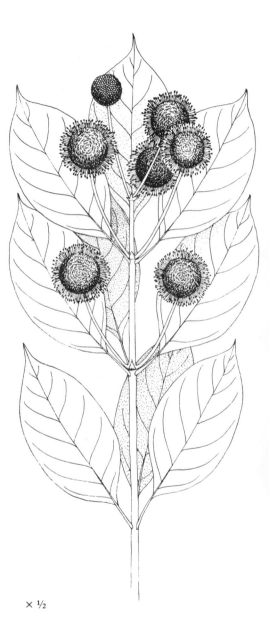

× ½

230

Dye Bedstraw or Clayton's Bedstraw
Galium tinctorium L.

Madder Family
Rubiaceae

Description. Low to medium-height, weakly erect or matted herbaceous plant, up to 2 feet tall or long; stems four-angled, sharp recurved teeth or prickles on angles, and much branched; simple, entire, oblong to lance-shaped leaves (⅕–⅘ inch long) with rough margins and midveins, arranged in whorls of usually five or six on main stem and of two to four on branches, leaves of single whorl often differing in size; small greenish white three-lobed flowers with smooth, short, and straight stalks (pedicels) borne in clusters of threes.

Flowering period. June through September.

Habitat. Irregularly flooded tidal fresh marshes; inland marshes, bogs, and swamps.

Range. Quebec and Newfoundland to Washington, south to North Carolina, Texas, and Arizona.

Similar species. Other bedstraws associated with tidal fresh marshes include *Galium palustre* (Marsh Bedstraw), *G. trifidum* (Small Bedstraw), and *G. triflorum* (Sweet-scented Bedstraw). *G. palustre* has many-flowered inflorescences and two to six leaves in whorls. *G. trifidum* has four leaves per whorl and flowers borne on long, curved pedicels. *G. triflorum* has six leaves per whorl, but its leaves are bristle-tipped with rough hairy margins.

× 1

Common Elder or Common Elderberry

Sambucus canadensis L.

Honeysuckle Family
Caprifoliaceae

Description. Deciduous shrub, up to 12 feet tall; multiple stems usually about 1 inch in diameter with thick, soft white center (pith) and light brown bark with numerous large raised bumps (lenticels); oppositely arranged, compound leaves divided into five to eleven, usually seven, fine-toothed, lance-shaped, stalked leaflets tapering to a distinct point distally, lower leaflets sometimes divided into three parts; numerous small white five-lobed tubular flowers borne on dense, somewhat flat-topped terminal inflorescences (cymes) with five spreading branches from end of twig; fruit dark purplish berry.

Flowering period. June through July.

Fruiting period. Middle to late summer.

Habitat. Irregularly flooded tidal fresh marshes and swamps; inland marshes, meadows, swamps, old fields, moist woods, and roadsides.

Range. Nova Scotia to Manitoba and South Dakota, south to Florida and Texas.

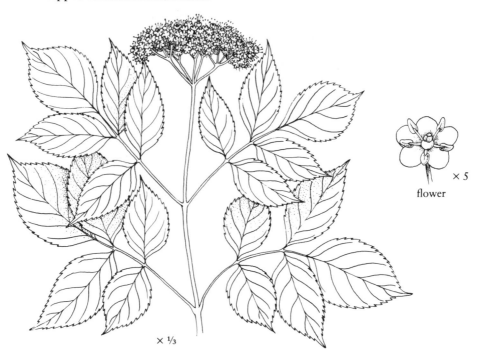

× 5
flower

× 1/3

Northern Arrowwood or Smooth Arrowwood

Viburnum recognitum Fernald

Honeysuckle Family
Caprifoliaceae

Description. Deciduous shrub, up to 15 feet tall; stems with smooth gray bark and white pith; twigs brown hairless, somewhat angled, up to four to six sides, new twigs light brown and distinctly angled; simple, coarse-toothed, egg-shaped to round leaves (1–3⅗ inches long and slightly less broad) tapering to a short point distally and slightly rounded at base, oppositely arranged; numerous small white five-lobed flowers in flat-topped or somewhat rounded inflorescences (cymes) at end of twigs; fruit dark blue, berrylike (drupe).

Flowering period. May through July.

Fruiting period. Summer.

Habitat. Irregularly flooded tidal fresh marshes and swamps; inland swamps and marshes, moist woods, and various drier sites.

Range. New Brunswick to Ontario, south to New Jersey, west to Michigan, and south in mountains only to Georgia.

Similar species. Viburnum dentatum (Southern Arrowwood) is the common arrowwood along the coastal plain from New Jersey south to Florida and Texas. Its twigs are velvety hairy, and its leaves are sometimes velvety hairy beneath. The leaves of *V. recognitum* are generally smooth beneath, except for hairs on veins. Other viburnums may be found in tidal fresh wetlands, such as *V. lentago* (Nannyberry), but they have fine-toothed leaves.

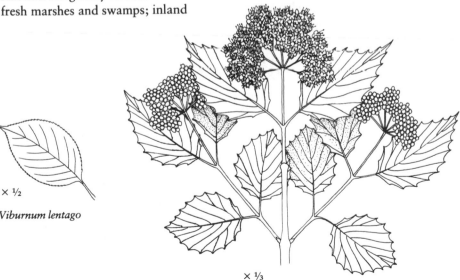

× ½

Viburnum lentago

× ⅓

233

Cardinal Flower
Lobelia cardinalis L.

Bluebell Family
Campanulaceae

Description. Medium-height to tall, erect herbaceous plant, 1½–5 feet high; perennial; stem typically unbranched, smooth or slightly hairy; simple, fine- or round-toothed, lance-shaped to oblong leaves (2–6 inches long and up to 2 inches wide) tapering at both ends, smooth or fine hairy, lower leaves short-stalked, upper leaves sessile, alternately arranged; numerous bright red (sometimes white) two-lipped tubular flowers (up to 1⅘ inches long) borne on terminal spikelike inflorescences (racemes), sometimes in leaf axils, upper lip of flower two-lobed and erect, lower lip three-lobed and spreading downward; fruit two-celled capsule.

Flowering period. July into October.

Habitat. Irregularly flooded tidal fresh marshes and swamps; inland marshes, swamps, and river banks.

Range. New Brunswick to Michigan and Minnesota, south into Florida and Texas.

× ⅔

Giant Ragweed
Ambrosia trifida L.

Composite or Aster Family
Compositae (Asteraceae)

Description. Medium-height to tall, erect herbaceous plant up to 17 feet high (occasionally); annual; stems hairy above and smooth below; simple, sharply three-to-five-lobed, toothed leaves (up to 8 inches long), rough on both sides, oppositely arranged; small green flowers in heads on dense spikes.

Flowering period. Late June into October.

Habitat. Tidal fresh marshes; river banks, moist soils, and waste places.

Range. Quebec to British Columbia, south to Florida and northern Mexico.

× ½

New York Aster
Aster novi-belgii L.

Composite or Aster Family
Compositae (Asteraceae)

Description. Medium-height to tall, erect herbaceous plant, 1½–5 feet high; perennial; stems sometimes with lines of hairs from leaf bases, sometimes smooth, except under flower heads; simple, entire or weakly toothed, narrowly lance-shaped leaves (1½–6¾ inches long) slightly clasping stem, alternately arranged; violet or blue daisylike flowers in heads (¾–1¼ inches wide) with twenty to fifty petallike rays (⅕–½ inch long) arranged in open or leafy inflorescences.

Flowering period. Late July into October.

Habitat. Slightly brackish and tidal fresh marshes, occasionally borders of salt marshes; inland marshes, shrub swamps, shores, and other moist areas.

Range. Newfoundland and Nova Scotia south to Georgia, apparently to Alabama, chiefly near the coast.

Similar species. Other asters of brackish marshes (*Aster tenuifolius* and *A. subulatus*) have somewhat fleshy, entire, linear leaves. In tidal fresh marshes, *A. puniceus* (Swamp Aster) and *A. simplex* (Lowland White Aster) also are present. *A. puniceus* usually has bluish daisylike flowers, sessile, somewhat clasping, usually toothed leaves with rough upper surfaces, and often hairy purplish stems (variable—

sometimes smooth green stems). *A. simplex* has weakly toothed or toothless (entire) petioled leaves and white, occasionally bluish daisylike flowers.

× 1

236

Bur Marigold

Bidens laevis (L.) B.S.P.

Composite or Aster Family
Compositae (Asteraceae)

Description. Medium-height, erect herbaceous plant, 1–3½ feet tall; annual or perennial; stems smooth; simple, toothed, sessile, lance-shaped leaves (2–6 inches long), midrib prominent, oppositely arranged; yellow daisylike flowers in small heads (1½–2½ inches wide) with seven to eight petallike yellow rays (⅘–1⅕ inches long); fruit barbed nutlet (achene).

Flowering period. August to November.

Habitat. Tidal fresh marshes; inland marshes and borders of ponds and streams.

Range. New Hampshire and Massachusetts to Florida and California, chiefly coastal, but also inland in north to Indiana and West Virginia.

Similar species. Bidens cernua (Nodding Beggar-ticks) and *B. hyperborea* (Northern Beggar-ticks) both have sessile leaves; *B. cernua* has rays less than ⅘ inch, sometimes absent, whereas *B. hyperborea* occurs in brackish marshes from Massachusetts north and has flowers in bell-shaped heads. Other *Bidens* in tidal fresh marshes (*B. eatoni, B. comosa, B. connata, B. frondosa, B. bipinnata, B. aristosa,* and *B. bidentoides*) have petioled leaves or deeply dissected leaves.

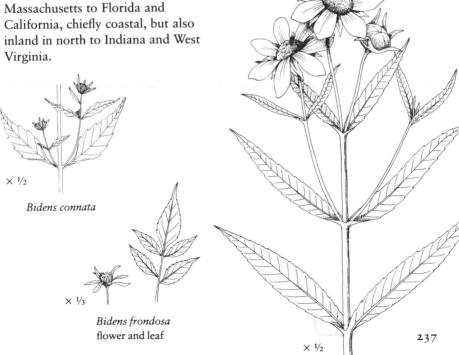

× ½

Bidens connata

× ⅓

Bidens frondosa
flower and leaf

× ½

237

Purple Joe-Pye-weed

Eupatoriadelphus purpureus (L.)
R. M. King & H. Rob.
(*Eupatorium purpureum* L.)

Composite or Aster Family
Compositae (Asteraceae)

Description. Medium-height to tall, erect herbaceous plant, 2–7 feet high; perennial; stems solid, somewhat waxy and green with purple at nodes; simple, coarsely toothed, lance-shaped to egg-shaped leaves (3 1/5–6 inches long), arranged in whorls of threes or fours; four to seven pale pink or purplish flowers in heads arranged in a round-topped terminal inflorescence.

Flowering period. Mid-July through September.

Habitat. Tidal fresh marshes; moist and dry thickets, inland marshes, and open woods.

Range. Southern New Hampshire to Minnesota and Nebraska, south to northern Florida and Oklahoma.

Similar species. Eupatoriadelphus maculatus (Spotted Joe-Pye-weed, formerly *Eupatorium maculatum*) has stem spotted with purple, leaves arranged in whorls of fours or fives, and purple flowers borne in a flat-topped inflorescence. *Eupatoriadelphus fistulosus* (Hollow-stemmed Joe-Pye-weed, formerly *Eupatorium fistulosum*) has a hollow, not solid, pith, and stem is more purplish throughout. *Eupatoriadelphus dubius* (Eastern

Joe-Pye-weed, formerly *Eupatorium dubium*) has a stem with purple spots, and its leaves have three main veins (one midvein and two lateral veins), whereas *E. purpureus* has one main vein. *Eupatorium perfoliatum* (Boneset) has clasping leaves and white flowers, and *Eupatorium serotinum* (Late Eupatorium) also has white flowers.

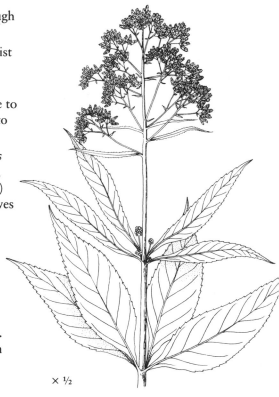

× 1/2

Boneset or Thoroughwort
Eupatorium perfoliatum L.

Composite or Aster Family
Compositae (Asteraceae)

Description. Medium-height to tall, erect herbaceous plant, 1½–5 feet high; perennial; stems with long spreading hairs, occasionally densely hairy or coarsely hairy; simple, coarsely toothed, triangle-shaped leaves (2¾–8 inches long) joined at bases to form a single leaf, oppositely arranged; nine to twenty-three white flowers in heads borne on a flat-topped inflorescence.

Flowering period. Late July through October.

Habitat. Tidal fresh marshes; inland marshes, wet meadows, shrub swamps, low woods, shores, and other moist areas.

Range. Nova Scotia and Quebec to Minnesota and Nebraska, south to Florida, Louisiana, and Texas.

Similar species. Other *Eupatorium* and *Eupatoriadelphus* do not have joined leaves. *Eupatorium serotinum* (Late Eupatorium) has white flowers and petioled leaves.

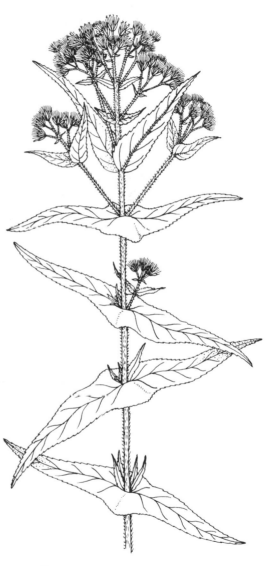

× 4

flower

× ½

239

Sneezeweed

Helenium autumnale L.

Composite or Aster Family
Compositae (Asteraceae)

Description. Medium-height to tall, erect herbaceous plant, 1–5 feet high; perennial; stems smooth or very fine hairy with wings; simple, shallowly toothed, lance-shaped leaves (2–6 inches long) forming wings along stem, alternately arranged; yellow daisylike flowers with ball-shaped head (disk) and ten to twenty wedge-shaped, drooping petallike rays having three-lobed broad tips, borne on long stalks (peduncles) in leafy inflorescence.

Flowering period. August through October.

Habitat. Tidal fresh marshes; inland marshes, wet meadows, shrub swamps, and shores.

Range. Quebec to Minnesota and Nebraska, south to Florida, Texas, and Arizona.

× ½

Climbing Hempweed
Mikania scandens (L.) Willd.

Composite or Aster Family
Compositae (Asteraceae)

Description. Twining and sprawling herbaceous vine, up to 20 feet long; stems hairy to nearly smooth; simple, slightly toothed leaves (1−5½ inches long), somewhat triangular-shaped with rounded bases or heart-shaped, tapering to a slender point, oppositely arranged; numerous white or pink flowers in heads borne in stalked clusters arising from leaf axils.

Flowering period. July through October.

Habitat. Tidal fresh marshes; inland marshes, swamps, and stream banks.

Range. Southern Maine to Florida and Texas; locally inland to Michigan and Missouri.

× ¾

New York Ironweed

Vernonia noveboracensis (L.) Michx.

Composite or Aster Family
Compositae (Asteraceae)

Description. Medium-height to tall, erect herbaceous plant, 3–7 feet high; stems smooth or thinly hairy; simple, finely toothed or nearly entire, lance-shaped or narrowly lance-shaped leaves (4–8 inches long), rough hairy above and thin hairy beneath, alternately arranged; twenty-nine to forty-seven purple flowers in heads (½–¾ inch wide) arranged in loose, open, flattened, or round-topped inflorescence.

Flowering period. August through October.

Habitat. Tidal fresh marshes; inland swamps, marshes, and stream banks, mostly near the coast.

Range. Massachusetts to Georgia and Mississippi, occasionally inland to Ohio and West Virginia.

× ½

Places to Observe Coastal Wetlands

Nearly all of us have observed coastal wetlands while driving to the beach along highways and secondary roads. For a closer look, however, I recommend a visit to a nearby refuge, wildlife management area, park, sanctuary, or nature center owned by federal and state agencies or by private nonprofit organizations, such as the National Audubon Society or the Nature Conservancy. These conservation areas have boardwalks across the marshes and/or hiking trails along wetland margins.

The following discussion briefly describes the distribution of coastal wetlands in each northeastern state and lists some places to walk in and about these wetlands. General locations of these sites are shown on state maps that also show the general distribution of coastal marshes. Other areas may also be available for public use. For this information, contact the appropriate local conservation commission, parks department, or planning office and various private environmental organizations (listed under Sources of Other Information).

Maine

Maine possesses about 135,000 acres of coastal wetlands. Most of the state's coastal marshes lie in the southwestern part of the coast from the Maine–New Hampshire border north to the Bath region. From the vicinity of Portland, Maine's renowned rocky shoreline begins, extending eastward to Lubec and Eastport. Here, coastal wetlands are generally limited to tidal flats and seaweed-covered rocky shores. Good places to see coastal wetlands include the following areas: (1) Vaughan Woods State Park, Eliot; (2) Rachel Carson National Wildlife Refuge (U.S. Fish and Wildlife Service), Wells; (3) Wells National Estuarine Sanctuary (Maine State Planning Office), Wells; (4) Scarborough Marsh Nature Center (Maine Audubon Society), Scarborough; (5) Wolf Neck State Park, Freeport; (6) Popham Beach State Park, Phippsburg; (7) Reid State Park, Georgetown; (8) Acadia National Park, Bar Harbor; (9) Petit Manan National Wildlife Refuge, Steuben; and (10) Cobscook Bay State Park, Edmunds.

New Hampshire

Although New Hampshire's coastline is small, it contains about 10,000 acres of coastal marshes and tidal flats. Driving along Route 1A, one can see salt marshes from Seabrook Beach to Rye North

Beach. The state's largest estuary—Great Bay—is situated west of Portsmouth and is part of the Piscataqua River system. Coastal wetlands border Great Bay and may be seen at Adams Point where the University of New Hampshire's Jackson Estuarine Laboratory is located.

Massachusetts

Coastal wetlands abound in Massachusetts. Roughly 90,000 acres of salt and brackish marshes and tidal flats are present. Salt marshes are best developed behind barrier islands and beaches, such as Plum Island, Sandy Neck, and Nauset Beach. Brackish marshes and tidal freshwater wetlands occur along large coastal rivers like the Merrimack and North rivers. The Cape Ann peninsula has seaweed-covered rocky shores, more typical of the Maine coastline. Coastal wetlands can be studied closely at the following locations: (1) Parker River National Wildlife Refuge (U.S. Fish and Wildlife Service), Newburyport; (2) Crane Reservation (The Trustees of Reservations), Ipswich; (3) Halibut Point Reservation (rocky shore only), Rockport; (4) World's End Reservation, Hingham; (5) Wellfleet Bay Wildlife Sanctuary (Massachusetts Audubon Society), South Wellfleet; (6) Cape Cod National Seashore (National Park Service), Eastham; (7) Cape Cod Museum of Natural History, Brewster; (8) Sandy Neck Reservation, Barnstable; (9) South Cape State Park, Mashpee; (10) Waquoit Bay National Estuarine Sanctuary (proposed; Massachusetts Coastal Zone Management Office), Falmouth; (11) Felix Neck Wildlife Sanctuary, Vineyard Haven; (12) Wasque Reservation, Chappaquiddick Island; (13) Lloyd Environmental Center, South Dartmouth; and (14) Horseneck Beach State Park, Westport.

Rhode Island

Although it is the smallest northeastern state, Rhode Island nonetheless has a variety of coastal wetlands, including about 7,000 acres of salt and brackish marshes, tidal flats, and rocky shores. The marshes are best developed along the state's salt ponds, such as Winnapaug Pond, Quonochontaug Pond, and Ninigret Pond. The following places are worth visiting: (1) Ninigret National Wildlife Refuge (U.S. Fish and Wildlife Service), Charlestown; (2) Trustom Pond National Wildlife Refuge, Green Hill; (3) Norman Bird Sanctuary (Audubon Society of Rhode Island), Middletown; and (4) Narragansett Bay National Estuarine Sanctuary (Rhode Island Department of Environmental Management), Prudence Island.

Connecticut

Coastal wetlands have formed along Long Island Sound behind barrier beaches and in semienclosed embayments. They are also extensive along large coastal rivers, such as the Connecticut, the Housatonic, and the Quinnipiac, where tidal freshwater wetlands develop. There are approximately 15,000 acres of tidal marshes in the state. A few places to see coastal wetlands include: (1) Barn Island Wildlife Management Area (Connecticut Department of Environmental Protection), Stonington; (2) Bluff Point Coastal Reserve (Connecticut Department of Environmental Protection), Groton; (3) Rocky Neck State Park, East Lyme; (4) Hammonasset Beach State Park and Natural Area, Madison and Clinton; (5) Milford Point Sanctuary (New Haven Bird Club), Milford; and (6) Sherwood Island State Park, Westport.

New York

Long Island has vast stretches of coastal marshes and tidal flats behind barrier islands, such as Fire Island and Jones Island, and within embayments along Long Island Sound. There are brackish and tidal freshwater wetlands along the Hudson River. Roughly 26,000 acres of salt and brackish marshes exist in New York. Numerous conservation areas are available for the public to visit coastal wetlands, including: (1) Jamaica Bay Wildlife Refuge (National Park Service), New York City; (2) Marine Nature Study Area (Town of Hempstead), Oceanside; (3) Jones Beach State Park, Nassau County; (4) Gilgo Beach State Park, Suffolk County;

(5) Fire Island National Seashore (National Park Service); (6) Wertheim National Wildlife Refuge (U.S. Fish and Wildlife Service), Brookhaven; (7) Merrill Lake Nature Trail (Accabonac Harbor Preserve), East Hampton; (8) Morton National Wildlife Refuge, Sag Harbor; (9) Mashomack Preserve (The Nature Conservancy), Shelter Island; (10) Montauk Point State Park, Montauk; (11) Orient Beach State Park, Orient; (12) Hubbard Creek Marsh (Suffolk County Park), Flanders; (13) Caumsett State Park, Lloyd Neck; (14) Welwyn Preserve (Nassau County Park), Glen Cove; (15) Marshlands Conservancy (Westchester County Park), Rye; (16) William T. Davis Wildlife Refuge, New Springville, Staten Island; and (17) Hudson River National Estuarine Sanctuary (New York State Department of Environmental Conservation): Piermont Marsh, Piermont, and Iona Island Marsh, Stony Point.

New Jersey

Of all the northeastern states, New Jersey has the greatest amount of coastal wetlands, with nearly 265,000 acres (excluding tidal swamps). An additional 31,000 acres of coastal aquatic beds exist in estuarine waters. Salt and brackish marshes are well developed behind barrier islands from Barnegat Bay to Cape May and along Delaware Bay. Tidal fresh marshes have formed along the Delaware River and other large coastal rivers, such as the Great Egg Harbor, the Mullica, and the Maurice. Coastal wetlands can be seen at the following areas: (1) Hackensack Meadowlands Environmental Center, Lyndhurst; (2) Cheesequake State Park, Cheesequake; (3) Gateway National Recreation Area (National Park Service), Sandy Hook; (4) Island Beach State Park, Seaside Park; (5) Tuckerton Wildlife Management Area (New Jersey Department of Environmental Protection), Tuckerton; (6) Forsythe National Wildlife Refuge (U.S. Fish and Wildlife Service)—Brigantine Division, Oceanville, and Barnegat Division, Barnegat; (7) Bass River State Forest, New Gretna; (8) Wetlands Institute, Stone Harbor; (9) Supawna Meadows National Wildlife Refuge, Salem; and (10) Rancocas State Park, Timbuctoo.

Pennsylvania

Only tidal freshwater wetlands exist in Pennsylvania's coastal zone along the Delaware River. Most of these wetlands are tidal flats, with a few coastal marshes also present. The best example of the tidal fresh marshes can be seen at Tinicum Environmental Center (U.S. Fish and Wildlife Service) in Philadelphia across from the International Airport.

Delaware

Most of Delaware's coastal wetlands have developed along Delaware Bay.

Numerous tidal wetlands are also associated with the inland bays in the southeastern part of the state—Rehoboth, Indian River, and Little Assawoman bays. There are freshwater tidal wetlands along the upper reaches of coastal rivers, such as the Christina, Smyrna, Leipsic, Murderkill, and Milton rivers that flow into Delaware Bay and the Nanticoke River that drains to the west into Chesapeake Bay. There are about 90,000 acres of salt and brackish marshes and tidal flats, with another 8,000 acres of tidal freshwater wetlands also present. The following areas provide good opportunities to view coastal wetlands: (1) Augustine Wildlife Area (Delaware Department of Natural Resources and Environmental Control), Port Penn; (2) Woodland Beach Wildlife Area, Woodland Beach; (3) Bombay Hook National Wildlife Refuge (U.S. Fish and Wildlife Service), Smyrna; (4) Little Creek Wildlife Area, Little Creek; (5) Prime Hook National Wildlife Refuge and Prime Hook Wildlife Area, Milton; (6) Cape Henlopen State Park, Lewes; (7) Delaware Seashore State Park, Dewey Beach; (8) Assawoman Wildlife Area, near Bethany Beach; and (9) Fenwick Island State Park, Fenwick Island.

Maryland

Chesapeake Bay and its tributaries dominate Maryland's coastal zone, creating vast acreages of salt, brackish, and tidal freshwater wetlands. Brackish marshes form along the shores of the upper bay and extend upstream some distance in coastal rivers. Tidal fresh marshes are associated with the larger coastal rivers, such as the Pocomoke, Chester, Choptank, and Nanticoke rivers on the eastern shore and the Patuxent and Potomac rivers on the western shore. Salt marshes are prevalent in the lower bay and along the Atlantic coast behind Assateague Island and Ocean City. Coastal wetlands in estuarine waters amount to roughly 170,000 acres, and there are about 45,000 acres of tidal freshwater marshes and swamps. Approximately 42,000 acres of coastal aquatic beds are present. Good places to see coastal wetlands include: (1) Elk Neck State Park, North East; (2) Sandy Point State Park, Annapolis; (3) Eastern Neck National Wildlife Refuge (U.S. Fish and Wildlife Service), Rock Hall; (4) Jug Bay Natural Area (Maryland–National Capitol Park and Planning Commission), Upper Marlboro; (5) Calvert Cliffs State Park, Calvert County; (6) Point Lookout State Park, Scotland; (7) Blackwater National Wildlife Refuge, Cambridge; (8) Deal Island Wildlife Management Area (Maryland Department of Natural Resources), Dames Quarter and Chesapeake Bay National Estuarine Sanctuary (Maryland Department of Natural Resources)—Monie Bay component, Princess Anne; (9) Irish Grove Sanctuary (Maryland Ornithological Society), Rumbly Point; and (10) Assateague Island National Seashore (National Park Service), Berlin (southern access through Chincoteague National Wildlife Refuge, Chincoteague, Virginia, which also has extensive coastal marshes open to the public).

Maine

1. Cobscook Bay State Park, Edmunds
2. Petit Manan National Wildlife Refuge, Steuben
3. Acadia National Park, Bar Harbor
4. Reid State Park, Georgetown
5. Popham Beach State Park, Phippsburg
6. Wolf Neck State Park, Freeport
7. Scarborough Marsh Nature Center, Scarborough
8. Wells National Estuarine Sanctuary, Wells
9. Rachel Carson National Wildlife Refuge, Wells
10. Vaughan Woods State Park, Eliot

New Hampshire

11. Jackson Estuarine Laboratory, Adams Point

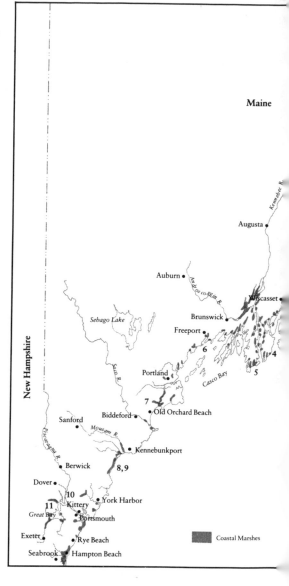

General distribution of coastal marshes in Maine and New Hampshire showing some wetland areas with good public access.

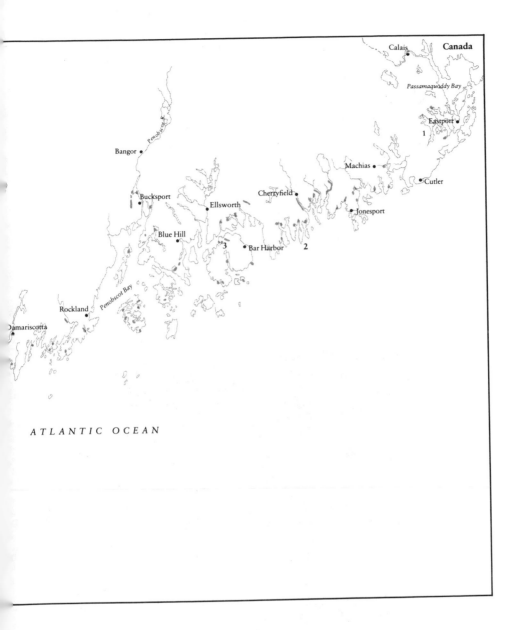

ATLANTIC OCEAN

Canada

Calais

Passamaquoddy Bay

Eastport

1

Bangor

Machias

Cutler

Bucksport

Cherryfield

Jonesport

Ellsworth

Blue Hill

3

Bar Harbor

2

Rockland

Penobscot Bay

Damariscotta

Penobscot R.

Massachusetts

1. Parker River National Wildlife Refuge, Newburyport
2. Crane Reservation, Ipswich
3. Halibut Point Reservation, Rockport
4. World's End Reservation, Hingham
5. Wellfleet Bay Wildlife Sanctuary, South Wellfleet
6. Cape Cod National Seashore, Eastham
7. Cape Cod Museum of Natural History, Brewster
8. Sandy Neck Reservation, Barnstable
9. South Cape State Park, Mashpee
10. Waquoit Bay National Estuarine Sanctuary, Falmouth
11. Felix Neck Wildlife Sanctuary, Vineyard Haven
12. Wasque Reservation, Chappaquiddick Island
13. Lloyd Environmental Center, South Dartmouth
14. Horseneck Beach State Park, Westport

Rhode Island

15. Norman Bird Sanctuary, Middletown
16. Narragansett Bay National Estuarine Sanctuary, Prudence Island
17. Trustom Pond National Wildlife Refuge, Green Hill
18. Ninigret National Wildlife Refuge, Charlestown

Buttonbush

New Hampshire

Merrimack R.

Newburyport

1

Lawrence

Ipswich

2

3

Gloucester

Salem

Massachusetts

Boston

Massachusetts Bay

Quincy

4

Weymouth

ATLANTIC OCEAN

Pembroke

Provincetown

Plymouth

Wellfleet

5

Cape Cod Bay

6

Orleans

Providence

Connecticut

Rhode Island

8

7

East Greenwich

Fall River

Hyannis

Chatham

16

New Bedford

13

Falmouth

10 9

Buzzards Bay

Kingston

15

11

Nantucket Sound

14

Newport

Westerly

Charlestown

18

17

Edgartown

12

Rhode Island Sound

Block Island Sound

Nantucket

Block Island

Coastal Marshes

General distribution of coastal marshes in
Massachusetts and Rhode Island showing
some wetland areas with good public access.

Connecticut

1. Barn Island Wildlife Management Area,
   Stonington
2. Bluff Point Coastal Reserve, Groton
3. Rocky Neck State Park, East Lyme
4. Hammonasset Beach State Park and
   Natural Area, Madison and Clinton
5. Milford Point Sanctuary, Milford
6. Sherwood Island State Park, Westport

Blue Flag

General distribution of coastal marshes in
Connecticut showing some wetland areas
with good public access.

New York

1. Jamaica Bay Wildlife Refuge,
   New York City
2. Marine Nature Study Area, Oceanside
3. Jones Beach State Park, Nassau County
4. Gilgo Beach State Park, Suffolk County
5. Fire Island National Seashore,
   Suffolk County
6. Wertheim National Wildlife Refuge,
   Brookhaven
7. Merrill Lake Nature Trail,
   East Hampton
8. Morton National Wildlife Refuge,
   Sag Harbor
9. Mashomack Preserve, Shelter Island
10. Montauk Point State Park, Montauk
11. Orient Beach State Park, Orient
12. Hubbard Creek Marsh, Flanders
13. Caumsett State Park, Lloyd Neck
14. Welwyn Preserve, Glen Cove
15. Marshlands Conservancy, Rye
16. William T. Davis Wildlife Refuge,
    New Springville
17. Hudson River National Estuarine
    Sanctuary, Piermont Marsh, Piermont
18. Hudson River National Estuarine
    Sanctuary, Iona Island Marsh,
    Stony Point

Rose Mallow

General distribution of coastal marshes in
New York showing some wetland areas with
good public access.

New Jersey

1. Hackensack Meadowlands
   Environmental Center, Lyndhurst
2. Cheesequake State Park, Cheesequake
3. Gateway National Recreation Area,
   Sandy Hook
4. Island Beach State Park, Seaside Park
5. Forsythe National Wildlife Refuge,
   Barnegat Division, Barnegat
6. Tuckerton Wildlife Management Area,
   Tuckerton
7. Forsythe National Wildlife Refuge,
   Brigantine Division, Oceanville

8. Bass River State Forest, New Gretna
9. Wetlands Institute, Stone Harbor
10. Supawna Meadows National Wildlife
    Refuge, Salem
11. Rancocas State Park, Timbuctoo

Pennsylvania

12. Tinicum Environmental Center,
    Philadelphia

Joe-Pye-weed

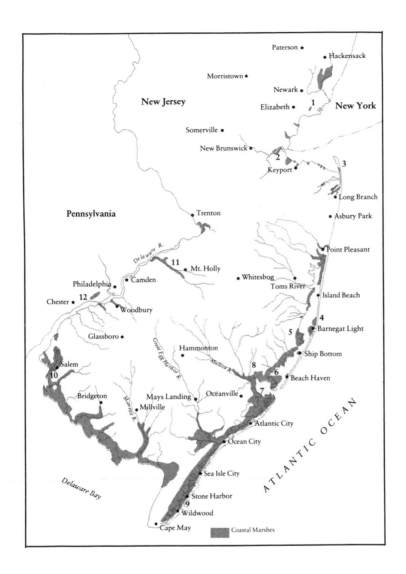

General distribution of coastal marshes in
New Jersey and Pennsylvania showing some
wetland areas with good public access.

Maryland

1. Elk Neck State Park, North East
2. Sandy Point State Park, Annapolis
3. Eastern Neck National Wildlife Refuge, Rock Hall
4. Jug Bay Natural Area, Upper Marlboro
5. Calvert Cliffs State Park, Calvert County
6. Point Lookout State Park, Scotland
7. Blackwater National Wildlife Refuge, Cambridge
8. Deal Island Wildlife Management Area and Chesapeake Bay National Estuarine Sanctuary (Monie Bay component), Princess Anne
9. Irish Grove Sanctuary, Rumbly Point
10. Assateague Island National Seashore, Berlin

Delaware

11. Augustine Wildlife Area, Port Penn
12. Woodland Beach Wildlife Area, Woodland Beach
13. Bombay Hook National Wildlife Refuge, Smyrna
14. Little Creek Wildlife Area, Little Creek
15. Prime Hook National Wildlife Refuge and Prime Hook Wildlife Area, Milton
16. Cape Henlopen State Park, Lewes
17. Delaware Seashore State Park, Dewey Beach
18. Assawoman Wildlife Area, Bethany Beach
19. Fenwick Island State Park, Fenwick Island

General distribution of coastal marshes in
Delaware and Maryland showing some
wetland areas with good public access.

Sources of Other Information

Those of you interested in learning more about coastal wetlands are referred to the following books. Other, more technical, scientific articles are presented in the References. Also listed below are agencies and organizations that may have other information (maps and reports) on coastal wetlands available for distribution. Special notations have been made for those agencies producing wetland maps.

Additional Readings

Life and Death of the Salt Marsh by John and Mildred Teal. 1969. Ballantine Books, New York, NY. (Nontechnical)

Wetlands of the United States: Current Status and Recent Trends by R. W. Tiner, Jr. 1984. U.S. Fish and Wildlife Service. U.S. Government Printing Office, Washington, DC. (Nontechnical)

Salt Marshes and Salt Deserts of the World by V. J. Chapman. 1960. Interscience Publishers, New York, NY. (Technical)

Ecology of Halophytes by R. J. Reimold and W. H. Queen (editors). 1974. Academic Press, New York, NY. (Technical)

The Ecology of a Salt Marsh by L. R. Pomeroy and R. G. Wiegert (editors). 1981. Springer-Verlag, New York, NY. (Technical)

The Ecology of New England High Salt Marshes: A Community Profile by S. W. Nixon. 1982. U.S. Fish and

Wildlife Service, Washington, DC. FWS/OBS-81/55. (Technical)

The Ecology of New England Tidal Flats: A Community Profile by R. B. Whitlatch. 1982. U.S. Fish and Wildlife Service, Washington, DC. FWS/OBS-81/01. (Technical)

Animals of the Tidal Marsh by F. C. Daiber. 1982. Van Nostrand Reinhold Co., New York, NY. (Technical)

The Ecology of Tidal Freshwater Marshes of the United States East Coast: A Community Profile by W. E. Odum, T. J. Smith III, J. K. Hoover, and C. C. McIvor. 1984. U.S. Fish and Wildlife Service, Washington, DC. FWS/OBS-83/17. (Technical)

Conservation of Tidal Marshes by F. C. Daiber. 1986. Van Nostrand Reinhold Co., New York, NY. (Technical)

Federal Agencies

U.S. Army Corps of Engineers
New England Division
424 Trapelo Road
Waltham, MA 02154

U.S. Army Corps of Engineers
New York District
26 Federal Plaza
New York, NY 10278

U.S. Army Corps of Engineers
Philadelphia District
2nd and Chestnut Streets
Philadelphia, PA 19106

U.S. Environmental Protection Agency
Region I
John F. Kennedy Federal Building
Boston, MA 02203

U.S. Environmental Protection Agency
Region II
26 Federal Plaza
New York, NY 10007

U.S. Environmental Protection Agency
Region III
841 Chestnut Building
Philadelphia, PA 19107

U.S. Fish and Wildlife Service
Region 5
1 Gateway Center
Newton Corner, MA 02158
(National Wetlands Inventory maps)

National Marine Fisheries Service
Northeast Regional Office
Federal Building
14 Elm Street
Gloucester, MA 01930

National Oceanic and Atmospheric
    Administration
National Ocean Service
Office of Ocean and Coastal Resource
    Management
3300 Whitehaven Street, NW
Washington, DC 20235
(coordinates National Estuarine
    Sanctuary Program)

National Park Service
North Atlantic Regional Office
15 State Street
Boston, MA 02109

National Park Service
Mid-Atlantic Regional Office
143 South 3rd Street
Philadelphia, PA 19106

State Agencies

Connecticut

Department of Environmental
    Protection
State Office Building
165 Capitol Avenue

Hartford, CT 06106
(coastal wetland maps)

Delaware

Department of Natural Resources and
    Environmental Control
Wetlands and Underwater Lands Branch
P.O. Box 1401
89 Kings Highway
Dover, DE 19903
(coastal wetland maps)

Maine

Department of Environmental
    Protection
State House
Augusta, ME 04333

Department of Marine Resources
State House
Augusta, ME 04333

Maryland

Department of Natural Resources
Tidewater Administration
Tawes State Office Building
Annapolis, MD 21401

Department of Natural Resources
Water Resources Administration
Tawes State Office Building
Annapolis, MD 21401
(coastal wetland maps)

Massachusetts

Coastal Zone Management Office
100 Cambridge Street
Boston, MA 02202

Department of Fisheries, Wildlife, and
    Recreational Vehicles
100 Cambridge Street
Boston, MA 02202

Department of Environmental Quality
    Engineering
Wetlands Division

1 Winter Street
Boston, MA 02108

New Hampshire

Fish and Game Department
34 Bridge Street
Concord, NH 03301

Wetlands Board
37½ Pleasant Street
Concord, NH 03301

New Jersey

Department of Environmental
  Protection
Division of Coastal Resources
Wetlands Section
CN-401
Trenton, NJ 08625
(coastal wetland maps)

New York

New York State Department of
  Environmental Conservation
Division of Marine Resources
Bureau of Marine Habitat Protection
State University of New York
Building 40
Stony Brook, NY 11794
(coastal wetland maps)

Pennsylvania

Department of Environmental Resources
Coastal Zone Management
P.O. Box 1467
Harrisburg, PA 17120

Rhode Island

Department of Environmental
  Management
Division of Coastal Resources
83 Park Street
Providence, RI 02903
(coastal wetland maps)

Private Environmental Organizations

National Organizations

American Littoral Society
Sandy Hook
Highlands, NJ 07732

National Audubon Society
950 3rd Avenue
New York, NY 10022

National Wildlife Federation
1412 16th Street, NW
Washington, DC 20036

The Nature Conservancy
Eastern Regional Office
294 Washington Street
Boston, MA 02181

The Sierra Club
Northeast Office
37–39 Trumbell Street, Suite 104
New Haven, CT 06511

State and Local Groups

Association of New Jersey
  Environmental Commissions
Box 157
Mendham, NJ 07945

Audubon Society of New Hampshire
3 Silk Farm Road
P.O. Box 528B
Concord, NH 03301

Audubon Society of Rhode Island
40 Bowen Street
Providence, RI 02903

Chesapeake Bay Foundation, Inc.
162 Prince George Street
Annapolis, MD 21401

Connecticut Audubon Society
2325 Burr Street
Fairfield, CT 06430

Delaware Nature Education Society, Inc.
Box 700
Hockessin, DE 19707

Delaware Wild Lands, Inc.
Fifth and Main Streets
Odessa, DE 19730

Maine Association of Conservation
   Commissions
125 Auburn Street
Portland, ME 04103

Maine Audubon Society
Gilsland Farm
118 Route 1
Falmouth, ME 04105

Marine Environmental Council of Long
   Island, Inc.
P.O. Box 55
Seaford, Long Island, NY 11783

Massachusetts Association of
   Conservation Commissions
Lincoln Filene Center
Tufts University
Medford, MA 02155

Massachusetts Audubon Society
South Great Road
Lincoln, MA 01773

New Hampshire Association of
   Conservation Commissions
54 Portsmouth Street
Concord, NH 03301

New Jersey Audubon Society
790 Ewing Avenue
Franklin Lakes, NJ 07417

New Jersey Conservation Foundation
300 Mendham Road
Morristown, NJ 07960

Save the Bay, Inc.
154 Francis Street
Providence, RI 02903

The Trustees of Reservations
224 Adams Street
Milton, MA 02186

References

Anderson, R. R., R. G. Brown, and R. D. Rappleye. 1968. Water quality and plant distribution along the upper Patuxent River, Maryland. *Chesapeake Science* 9: 145–156.

Brown, M. L., and R. G. Brown. 1984. *Herbaceous Plants of Maryland.* University of Maryland Book Store, College Park.

Brown, R. G., and M. L. Brown. 1972. *Woody Plants of Maryland.* University of Maryland Book Store, College Park.

Carter, V., P. T. Gammon, and N. C. Bartow. 1983. *Submerged Aquatic Plants of the Tidal Potomac River.* U.S. Geological Survey Bulletin 1543.

Chapman, V. J. 1960. *Salt Marshes and Salt Deserts of the World.* Interscience Publishers, New York.

Cobb, B. 1963. *A Field Guide to the Ferns and Their Related Families of Northeastern and Central North America.* Houghton Mifflin Co., Boston.

Conrad, H. S. 1935. The plant associations of central Long Island. *American Midland Naturalist* 16: 433–516.

Cowardin, L. W., V. Carter, F. C. Golet, and E. T. LaRoe. 1979. *Classification of Wetlands and Deepwater Habitats of the United States.* U.S. Fish and Wildlife Service, Washington, DC. FWS/OBS-79/31.

Fassett, N. C. 1966. *A Manual of Aquatic Plants.* University of Wisconsin Press, Madison.

Fefer, S. I., and P. A. Schettig. 1980. *An Ecological Characterization of Coastal Maine (North and East of Cape Elizabeth).* Volume 4. Appendixes. U.S. Fish and Wildlife Service, Newton Corner, MA. FWS/OBS-80/29.

Fender, F. S. 1938. The flora of Seven Mile Beach, New Jersey. *Bartonia* 19: 23–41.

Fernald, M. L. 1970. *Gray's Manual of Botany.* 8th (Centennial) edition. D. Van Nostrand Co., New York.

Ferren, W. R., Jr. 1976. Aspects of intertidal zones, vegetation, and flora of the Maurice River system, New Jersey. *Bartonia* 44: 58–67.

Ferren, W. R., Jr., R. E. Good, R. Walker, and J. Arsenault. 1981. Vegetation and flora of Hog Island, a brackish wetland in the Mullica River, New Jersey. *Bartonia* 48: 1–10.

Ferren, W. R., Jr., and A. E. Schuyler. 1980. Intertidal vascular plants of river systems near Philadelphia. *Proceedings of the Academy of Natural Sciences of Philadelphia* 132: 86–120.

Flowers, M. G. 1973. Vegetation zonation in two successional brackish marshes of the Chesapeake Bay. *Chesapeake Science* 14: 197–200.

Gleason, H. A. 1952. *The New Britton and Brown Illustrated Flora of the Northeastern United States and Adjacent Canada.* 3 volumes. Hafner Press, Macmillan Publishing Co., New York.

Gleason, H. A., and A. Cronquist. 1963. *Manual of Vascular Plants of Northeastern United States and Adjacent Canada.* D. Van Nostrand Co., New York.

Good, R. E. 1965. Salt marsh vegetation, Cape May, New Jersey. *Bulletin of the New Jersey Academy of Science* 10: 1–11.

Good, R. E., and N. F. Good. 1975. Vegetation and production of the Woodbury Creek–Hessian Run freshwater tidal marshes. *Bartonia* 43: 38–45.

Hankin, A. L., L. Constantine, and S. Bliven. 1985. *Barrier Beaches, Salt Marshes, and Tidal Flats: An Inventory of the Coastal Resources of the Commonwealth of Massachusetts.* Massachusetts Coastal Zone Management Program. Publication 13899-27-600-1-85 C.R.

Hitchcock, A. S. 1950. *Manual of the Grasses of the United States.* U.S. Department of Agriculture, Washington, DC. Miscellaneous Publication no. 200.

Kennard, W. C., W. M. Lefor, and D. L. Civco. 1983. *Analysis of Coastal Marsh Ecosystems: Effect of Tides on Vegetational Change.* University of Connecticut, Institute of Water Resources, Storrs.

Knobel, E. 1977. *Field Guide to the Grasses, Sedges, and Rushes of the United States.* Dover Publications, New York. Reprint.

Krauss, R. W., R. G. Brown, R. D. Rappleye, A. B. Owens, C. Shearer, E. Hsiao, and J. L. Reveal. 1971. *Check List of Plant Species of the Chesapeake Bay Occurring within the Hightide Limits of the Bay and Its Tributaries.* University of Maryland, Dept. of Botany, College Park. Technical Bulletin 2002.

Kulik, S., P. Salmansohn, M. Schmidt, and H. Welch. 1984. *The Audubon Society Field Guide to the Natural Places of the Northeast: Coastal.* Pantheon Books, New York.

Lawrence, S. 1984. *The Audubon Society Field Guide to the Natural Places of the Mid-Atlantic States: Coastal.* Pantheon Books, New York.

Lefor, M. W., D. L. Civco, and W. C. Kennard. 1981. Distribution of salt marsh vegetation in relation to tidal datums. *Proceedings of National Workshop on In-place Resource Inventories: Principles and Practices* (University of Maine, Orono, August 9–14, 1981), pp. 182–188.

Lefor, M. W., and R. W. Tiner. 1972. *Report of the Consultant Biologists for the Period December 22, 1969, to June 30, 1972.* Tidal Wetlands Survey of the State of Connecticut. University of Connecticut, Biological Sciences Group, Storrs.

———. 1974. *Report of the Consultant Biologists for the Period August 1, 1972–December 31, 1973.* Tidal Wetlands Survey of the State of Connecticut. University of Connecticut, Biological Sciences Group, Storrs.

Magee, D. W. 1981. *Freshwater Wetlands: A Guide to Common Indicator Plants of the Northeast.* University of Massachusetts Press, Amherst.

McCormick, J., and T. Ashbaugh. 1972. Vegetation of a section of Oldmans Creek tidal marsh and related areas in Salem and Gloucester Counties, New Jersey. *Bulletin of the New Jersey Academy of Science* 17: 31–37.

McCormick, J., and H. A. Somes, Jr. 1982. *The Coastal Wetlands of Maryland.* Maryland Department of Natural Resources, Coastal Zone Management Program, Annapolis.

Miller, W. B., and F. E. Egler. 1950. Vegetation of the Wequetequock–Pawcatuck tidal marshes, Connecticut. *Ecological Monographs* 20: 143–172.

National Wildlife Federation. 1984. *Conservation Directory 1984.* Washington, DC.

Newcomb, L. 1977. *Newcomb's Wildflower Guide.* Little, Brown and Co., Boston, MA.

Nichols, G. E. 1920. The vegetation of Connecticut. VII. The associations of depositing areas along the seacoast. *Bulletin of the Torrey Club* 47: 511–548.

Niering, W. A., and R. S. Warren. 1980. Vegetation patterns and processes in New England salt marshes. *BioScience* 30: 301–307.

Nixon, S. W. 1982. *The Ecology of New England High Salt Marshes: A Community Profile.* U.S. Fish and Wildlife Service, Washington, DC. FWS/OBS-81/55.

Nixon, S. W., and C. A. Oviatt. 1973. Ecology of a New England salt marsh. *Ecological Monographs* 43: 463–498.

O'Connor, J. S., and O. W. Terry. 1972. *The Marine Wetlands of Nassau and Suffolk Counties, New York.* Marine Sciences Research Center, State University of New York, Stony Brook.

Odum, W. E., T. J. Smith III, J. K. Hoover, and C. C. McIvor. 1984. *The Ecology of Tidal Freshwater Marshes of the United States East Coast: A Community Profile.* U.S. Fish and Wildlife Service, Washington, DC. FWS/OBS-83/17.

Philipp, C. C., and R. G. Brown. 1965. Ecological studies of transition-zone vascular plants in South River, Maryland. *Chesapeake Science* 6: 73–81.

Porter, C. L. 1967. *Taxonomy of Flowering Plants.* W. H. Freeman and Co., San Francisco.

Radford, A. E., H. E. Ahles, and C. R. Bell. 1973. *Manual of the Vascular Flora of the Carolinas.* University of North Carolina Press, Chapel Hill.

Reimold, R. J., and W. H. Queen (editors). 1974. *Ecology of Halophytes.* Academic Press, New York.

Simpson, R. L., R. E. Good, M. A. Leck, and D. F. Whigham. 1983. The ecology of freshwater tidal wetlands. *BioScience* 33: 255–259.

Taylor, N. 1938. A preliminary report on the salt marsh vegetation of Long Island, New York. *Bulletin of the New York State Museum* 316: 21–84.

Tiner, R. W., Jr. 1985a. *Wetlands of Delaware.* Cooperative publication. U.S. Fish and Wildlife Service, National Wetlands Inventory, Newton Corner, MA, and Delaware Department of Natural Resources and Environmental Control, Wetlands Section, Dover.

———. 1985b. *Wetlands of New Jersey.* U.S. Fish and Wildlife Service, National Wetlands Inventory, Newton Corner, MA.

U.S. Department of Agriculture, Soil Conservation Service. 1982. *Soil Survey of York County, Maine.*

U.S. Department of Commerce, National Oceanic and Atmospheric Administration, National Ocean Survey. 1978. *Tide Tables 1979 High and Low Water Predictions, East Coast of North and South America including Greenland.*

U.S. Fish and Wildlife Service. 1965a. *Supplementary Report on the Coastal Wetlands Inventory of Long Island, N.Y.* Boston, MA.

———. 1965b. *Supplementary Report on the Coastal Wetlands Inventory of New Hampshire.* Boston, MA.

Whitlatch, R. B. 1982. *The Ecology of New England Tidal Flats: A Community Profile.* U.S. Fish and Wildlife Service, Washington, DC. FWS/OBS-81/01.

Glossary

Achene. Small dry, hard, one-seeded nutlet.

Alternate (leaves). Arranged singly along the stem, alternating from one side to the other up the stem.

Angled (stem). Having distinct edges; three-angled (triangular in cross-section) and four-angled (square).

Annual. Plant living for only one year; propagates from seeds.

Anther. Distal end of a stamen where pollen is produced.

Appressed. Closely compacted together, as in an *appressed* inflorescence.

Aromatic. Sweet-smelling.

Arrowhead-shaped (leaves). Appearing like an arrowhead, triangular in shape.

Ascending. Rising upward and somewhat spreading, as in an *ascending* inflorescence.

Awn. Bristle-shaped appendage.

Axil. Angle formed by a leaf or branch with the stem.

Axillary. Located in an axil.

Basal (leaves). Arising directly from the roots; may ascend along stem as sheaths and appear alternately arranged, as in cattails.

Berry. Fleshy or pulpy fruit.

Blade. Flattened leaf.

Bract. Leaflike or modified appendage subtending a flower or belonging to an inflorescence.

Bristle. Long stiff, hairlike structure.

Bud. Unexpanded flower.

Callous. Fleshy, thickened tissue, as in the tips of leaf teeth in certain plants.

Calyx. Outermost parts of a flower; refers to the sepals, which are usually green but sometimes colored and petallike.

Capsule. Dry fruit composed of two or more cells or chambers.

Catkin. Scaly spike of inconspicuous flowers lacking petals.

Cell. One of the chambers of a capsule.

Channeled. Having distinct grooves or ridges.

Clasping (leaves). Closely surrounding the stem and attached directly without stalk.

Compound (leaves). Divided into two or more distinct, separate parts (leaflets).

Corm. Enlarged fleshy base of stem; bulblike.

Corolla. Petals of a flower.

Cyme. Flowering inflorescence with innermost or terminal flowers blooming first.

Deciduous. Not persistent, dropping off plant after completing its function, as with *deciduous* leaves in fall.

Decumbent. Reclining or prostrate at base, with the upper part erect or ascending, as in *decumbent* stems.

Dioecious. Having two types of flowers (male and female) borne on separate plants.

Disk. Tubular flower forming the central head of composites or asters.

Dissected. Deeply divided, often into threadlike parts, as in *dissected* leaves.

Drupe. Fleshy or pulpy fruit having a single stone or pit.

Emergent. Herbaceous (nonwoody) plant standing erect.

Entire (leaves). Having smooth margins, without teeth.

Evergreen. Persistent, as in *evergreen* leaves that remain on plant through winter.

Filament. Basal part of a stamen that supports the anther.

Fleshy. Soft, thickened tissue; succulent.

Frond. Leaf of a fern.

Gland. Secreting structure or organ.

Glandular. Bearing glands.

Glume. Thin bract at the base of a grass spikelet.

Grain. Fruit of certain grasses.

Head. Dense cluster of sessile or nearly sessile flowers, characteristic of composites or asters.

Herbaceous. Nonwoody.

Hood. Erect, outermost "petals" of milkweed flowers.

Horn. Erect, inner tubular structure of milkweed flowers.

Inflorescence. Flowering part of a plant.

Internode. Portion of a stem between two nodes.

Irregular (flower). Similar parts (e.g., petals) differing in size and/or shape.

Irregularly flooded. Flooded by tides less than once daily.

Jointed (stem). Having obvious nodes.

Lance-shaped (leaves). Appearing as the head of a lance, several times longer than wide, broadest just above the base, tapering to a tip.

Lateral. Borne on the sides of a plant.

Lemma. Lower of two bracts enclosing the flower of a grass.

Lenticel. Corky spot or line, sometimes raised, on the bark of many trees and shrubs.

Ligule. Membranous or hairy structure at the junction of the leaf blade and the leaf sheath in grasses.

Linear. Narrow and elongate, several to many times longer than wide.

Lip. Upper and lower parts of certain tubular flowers.

Lobe. Indented part of leaf or flower, not divided into distinct and separate parts but still interconnected to similar parts of leaf or flower (e.g., petal).

Midrib. Central, prominent rib or main vein of a leaf, usually in center of leaf.

Midvein. Middle vein of a leaf.

Nerve. Vein of a leaf, usually the more prominent ones.

Node. Point of a stem where leaves and branches are produced.

Nutlet. A small, dry, hard fruit.

Oblong (leaves). Longer than wide, with nearly parallel sides.

Ocrea. Tubular stipule, in smartweeds becoming fibrous.

Opposite (leaves). Arranged in pairs along the stem.

Orifice. Opening of a leaf sheath along the stem.

Oval. Broadly egg-shaped, widest in the middle and tapering to the ends.

Ovary. Part of a pistil containing the seeds.

Palea. Upper of two bracts enclosing the flower of a grass.

Panicle. Much branched flowering inflorescence.

Panne. Shallow depression within irregularly flooded salt marshes.

Pedicel. Stalk of a single flower in a cluster.

Peduncle. Primary stalk of a flowering cluster or single flower.

Perennial. Plant living for many years, usually supported by underground parts, e.g., rhizomes, corms, tubers, or bulbs.

Perigynium. Inflated sac enclosing the seed of a sedge.

Persistent. Remaining on plant after function ceases, as in *persistent* fruits.

Petiole. Stalk of a leaf.

Pinnate (leaves). Divided into leaflets that are oppositely arranged.

Pistil. Seed-bearing structure of a flower, usually consisting of an ovary, stigma, and style.

Pith. Soft, fleshy, or spongy center of a stem.

Plano-convex. Flattened but somewhat curved.

Pod. Dry fruit capsule.

Prickly. Bearing small spines.

Prostrate. Lying flat on the ground.

Raceme. Spikelike inflorescence with stalked flowers.

Rachis. Main axis of a spike, branching inflorescence, or compound leaf.

Ranks. Number of rows of organs, such as leaves, along a stem.

Ray. Outer flower of the flowering head of composites or asters, often petallike.

Recurved. Curved downward.

Regular (flower). Similar flower parts of the same size and shape, radially symmetrical.

Regularly flooded. Flooded by tides at least once a day.

Rhizome. Underground part of a stem, usually horizontal and rooting at nodes and producing erect stems.

Runner. Prostrate, slender aboveground stem producing new plants at nodes.

Samara. Winged dry fruit bearing one seed.

Scale. Modified leaf or thin flattened structure.

Scape. Naked flowering stalk arising directly from roots.

Sepal. Outermost part of a flower, usually green but sometimes colored and petallike.

Septa. Partitions.

Sessile. Without stalks, as in *sessile* leaves that are attached directly to the stem without stalks.

Sheath. Tubular envelope surrounding the stem, as in leaf *sheaths* of grasses and sedges.

Shrub. Erect, woody plant less than 20 feet tall, usually with multiple stems but also including saplings of tree species.

Simple (leaves). Not divided into separate parts; leaf blade continuous.

Sinus. Space between two lobes.

Sori. Cluster of fruit dots of ferns.

Spadix. Fleshy spike.

Spathe. Large bract or pair of bracts enclosing an inflorescence.

Spatulate. Spoon-shaped.

Spike. Simple, unbranched inflorescence composed of a central axis with sessile or nearly sessile flowers.

Spikelet. Secondary spike.

Spine. Sharp-pointed outgrowth of stem.

Sporangia. Spore cases of ferns, horsetails, and quillworts.

Spore. Reproductive organ of ferns, horsetails, and quillworts.

Spur. Hollow, tubular extension of a flower, usually bearing nectar.

Stamen. Pollen-bearing part of a flower.

Stigma. Part of a pistil receiving and germinating pollen.

Stipules. Pair of appendages at the base of a leaf stalk or on each side of its attachments to the stem.

Style. Part of a pistil connecting the stigma with the ovary.

Stolon. Prostrate, slender aboveground stem producing new plants at nodes.

Submerged. Underwater.

Subtended. Lying below.

Succulent. Fleshy.

Sword-shaped (leaves). Appearing bayonet-shaped, flattened and tapering to a sharp-pointed tip.

Synonym. Previous taxonomic or scientific name.

271

Taproot. Prominent, deep-penetrating root.

Tidal. Subject to influence of ocean-driven tides.

Tree. Woody plant greater than 20 feet tall, with a single main stem (trunk).

Tuber. Short, thickened, usually underground stem, having buds or eyes and storing food.

Twining. Climbing by wrapping around another plant or other support.

Umbel. Branched inflorescence with flowering stalks arising from a single point.

Valve. Piece of an open capsule.

Vascular. Having vessels or ducts.

Veins. Threads of vascular tissue in a leaf.

Whorl. Three or more organs arranged in a circle around the stem, as in *whorled* leaves.

Wing. Flattened expansion of an organ, as the continuation of a leaf as a *wing* along the stem.

Index

Page references in boldface refer to plant descriptions and illustrations.

ramosissimum, **126**
ramosissimum var. prolificum, 126
sagittatum, 19, 198, **200**
Polypodiaceae, 163, 164
Polypody Fern Family, 163, 164
Pondweed
 Baby, 21, 85, 86
 Bushy, 88, **90**
 Clasping-leaved, 21, **86**
 Curly, 21, **83**, 84, 86
 Family, 83–89
 Flat-stem, 85, 86
 Heartleaf, 84
 Horned, 21, 87, **88**, 90
 Largeleaf, 84
 Leafy, 21, 85, 86
 Longleaf, 84
 Ribbonleaf, 21, **84**
 Robbins', 85
 Sago, 11, 21, **85**, 86
Pontederiaceae, 187, 188
Pontederia cordata, 19, **188**, 202
Potamogeton
 amplifolius, 84
 crispus, 21, **83**, 84, 86
 epihydrus, 21, **84**
 foliosus, 21, 85, 86
 nodosus, 84
 pectinatus, 11, 21, **85**, 86
 perfoliatus, 21, **86**
 pulcher, 84
 pusillus, 21, 85, 86
 richardsonii, 86
 robbinsii, 85
 zosteriformis, 85, 86
Potamogetonacea, 83–89
Potentilla anserina, 13, **136**
Prairie Cordgrass, 13, 15, 110, **112**
Primrose Family, 142, 143
Primulaceae, 142, 143
Ptilimnium capillaceum, **141**
Puccinellia
 distans, 107
 fasciculata, 107
 maritima, **107**
 pumila, 107
Purple Gerardia, 148
Purple Joe-Pye-weed, **238**
Purple Loosestrife, 15, 19, **216**

Pygmyweed, 9, **204**

Quackgrass, Stiff-leaf, **101**
Quillwort
 Family, 162
 Riverbank, 9, **162**

Ragweed, Giant, 235
Ranunculaceae, 134, 203
Ranunculus
 cymbalaria, 9, **134**
 sceleratus, 134
 subrigidus, 134
 trichophyllus, 134
Red Ash, 221
Red Fescue, **104**
Redhead-grass, 21, **86**
Red Maple, 19, **211**
Red Milkweed, 222
Red Osier, 219
Reed Canary Grass, **173**
Reed, Common, 14, **106**
Rhododendron viscosum, 21
Rhus
 radicans, **209**
 vernix, 209
Rice
 Cutgrass, 19, **172**
 Wild, 19, **174**
Riverbank Quillwort, 9, 162
River Bulrush, 19, 119, 120, 121, 180
Rocket, Sea, 135
Rockweeds, 9, 11
Rocky shores, 7, 9
Rosa
 palustris, 19, **205**
 rugosa, 205
Rosaceae, 136, 205
Rose
 Family, 136, 205
 Rugosa, 205
 Swamp, 19, **205**
Rose Mallow, 13, 14, 137, **138**, 139
Rosemary, Marsh, **144**
Royal Fern, **165**
Royal Fern Family, 165
Rubiaceae, 230, 231
Rugosa Rose, 205

CONVERSION TABLE

English Units	Metric Equivalents
1/25 inch	1 millimeter
1/5 inch	5 millimeters
1/4 inch	6 millimeters
1/2 inch	12 millimeters
1 inch	2.5 centimeters
1 foot	30 centimeters
3.3 feet	1 meter
10 feet	3 meters

Measurement scale in inches with corresponding metric correlations